T5-ANV-854

**PERGAMON INTERNATIONAL LIBRARY**
of Science, Technology, Engineering and Social Studies,

*The 1000-volume original paperback library in aid of education,
industrial training and the enjoyment of leisure*

Publisher: Robert Maxwell, M.C.

---

# TECHNOLOGY AND SCIENCE IN THE PEOPLE'S REPUBLIC OF CHINA

## An Introduction

## THE PERGAMON TEXTBOOK INSPECTION COPY SERVICE

An inspection copy of any book published in the Pergamon International Library will gladly be sent to academic staff without obligation for their consideration for course adoption or recommendation. Copies may be retained for a period of 60 days from receipt and returned if not suitable. When a particular title is adopted or recommended for adoption for class use and the recommendation results in a sale of 12 or more copies, the inspection copy may be retained with our compliments. The Publishers will be pleased to receive suggestions for revised editions and new titles to be published in this important International Library.

## Other Titles of Interest

BALASSA, B.
Policy Reform in Developing Countries

BHALLA, A.
Towards Global Action for Appropriate Technology

COLE, S.
Global Models and the International Economic Order

COLE, S. and LUCAS, H.
Models, Planning and Basic Needs

EPSTEIN, T. S. and JACKSON, D.
The Feasibility of Fertility Planning

FITZGERALD, R.
Human Needs and Politics

MAXWELL, N.
China's Road to Development

RODZINSKI, W.
A History of China

SALAS, R.
International Population Assistance: The First Decade

SINHA, R. and DRABEK, A.
The World Food Problem: Consensus and Conflict

# TECHNOLOGY AND SCIENCE IN THE PEOPLE'S REPUBLIC OF CHINA

## An Introduction

by

### JON SIGURDSON
*Director, Research Policy Institute*
*University of Lund, Sweden*

## PERGAMON PRESS

OXFORD · NEW YORK · TORONTO · SYDNEY · PARIS · FRANKFURT

| U.K. | Pergamon Press Ltd., Headington Hill Hall, Oxford OX3 0BW, England |
| U.S.A. | Pergamon Press Inc., Maxwell House, Fairview Park, Elmsford, New York 10523, U.S.A. |
| CANADA | Pergamon of Canada, Suite 104, 150 Consumers Road, Willowdale, Ontario M2J 1P9, Canada |
| AUSTRALIA | Pergamon Press (Aust.) Pty. Ltd., PO Box 544, Potts Point, N.S.W. 2011, Australia |
| FRANCE | Pergamon Press SARL, 24 rue des Ecoles, 75240 Paris, Cedex 05, France |
| FEDERAL REPUBLIC OF GERMANY | Pergamon Press GmbH, 6242 Kronberg-Taunus, Pferdstrasse 1, Federal Republic of Germany |

Copyright © 1980 Jon Sigurdson

*All Rights Reserved. No part of this publication may be reproduced, stored in a retrieval system or transmitted in any form or by any means: electronic, electrostatic, magnetic tape, mechanical, photocopying, recording or otherwise, without permission in writing from the publishers*

First edition 1980

**British Library Cataloguing in Publication Data**

Sigurdson, Jon
Technology and science in the People's Republic of China.
1. Science – China
2. Technology – China
I. Title
509'.51      Q127.C5      79–40575

**ISBN 0-08-024288-X hardcover**

Q
127
C5
S56
1979
R2295

Printed in Great Britain by Biddles Ltd, Guildford, Surrey

# Contents

v

# Preface

THIS book grew out of a request by the Society for Anglo-Chinese Understanding to write a popular booklet on science and technology in China. The time of drafting coincided with the preparation of the National Science Conference, in spring 1978, a short visit to China at the same time, and an increasing amount of information on various aspects of research and planning being made available in China. Consequently the booklet grew into a book. Although there may still be many mistakes of fact or misunderstanding, I have benefited greatly from visits to China in 1973 and 1978, hosted on both occasions by the Chinese Academy of Sciences. At final proofreading in early November 1979 the author included some additional information on China's research structure and resources, and a few comments on changing priorities, released in the preceding months. The early drafts have been improved through the comments of numerous colleagues and close friends. In particular I want to thank Professor Richard Suttmeier, Hamilton College, USA, and Mr. Nils Ekblad, in charge of the Office of the Scientific and Technical Attache at the Swedish Embassy in Beijing and his assistant Ola Svensson, who have generously provided background material, comments, and Chinese-language material on a substantial scale. I am indebted to the China Photo Service, Beijing, for providing the photographs which appear in this volume.

# Introduction[1]

## The Modernization Process

The time may be particularly ripe for discussing technology and science in China, and the reasons are manifold. A debate — not limited only to technology and science, raging in China since the early 1970s but in particular after the First Session of the Fourth National People's Congress in early 1975 — has now produced clear and well-publicized policies which are in the process of being implemented. A striking characteristic of these new policies is the declared intention to expand rapidly the use of foreign technology and to collaborate with the advanced capitalist countries.

This is a striking departure from the policies of the Cultural Revolution, and the changes have their roots in two conditions. First, with a rapidly advancing technological front in almost all industrial sectors and heavily affecting military technologies — a development almost completely controlled by the already industrialized countries — China has felt an urgent need to use resources available in the global economic system. Second, the need to reorient her technological and scientific policies has been reinforced by the mismanagement of her research institutions and institutions of higher education, now attributed to the influence of the previous leadership.

Much information on the role and organization of research has been made available, so it is relatively easy to present, in some detail, the issues which have been hotly debated. China is now reorganizing and quickly developing her institutions of technology and science in order to make them better serve the country. In this process China has clearly indicated that increased technological and scientific contacts with other countries are important. The science and technology sector has also been designated as one of the "four modernizations" together with industry, agriculture, and defence.

This book represents an attempt to present certain important facts about technological development, the availability of resources, the role of technicians, and the build-up of institutions — all in order to show the emphasis given by the Chinese policy-makers to the development of a high-technology sector. In addition, the related problem of how a developing socialist country copes with the difficult issue of environmental protection is discussed in a separate chapter. However, those interested in the transferability of Chinese solutions and approaches will find relatively little of interest in this book although comparisons are made in many places. The obvious reason for this "neglect" is that Chinese solutions require the social and political structure of the People's Republic of China and the political will to act.

[1] China's State Council has decided to use the Chinese phonetic alphabet — *pinyin* — to standardize the romanization of Chinese names of persons and places. Accordingly all publications put out in China from the 1st of January 1979 started to use this alphabet; a practice which has been adopted in the final revision of the manuscript of this book. However, the names of overseas Chinese will be spelt according to the usual way. For a number of references used by the author it has not been possible to identify the Chinese characters on which the earlier romanization has been based. These uncertain names of places and persons are underlined in the text or notes.

1

The planners and politicians in today's China emphasize that the question of speed is an acute one in the country's socialist construction. Many of them apparently view China's construction of the material base in relation to the world's industrial giants — in particular the United States. In one of a series of articles in autumn 1977, the State Planning Commission referred to Mao's comparison in 1956 of the situation in China and the United States. Mao is now seen as having at the time set the goal of overtaking the United States economically in 50 or 60 years. Chairman Mao said:

> This is an obligation. You have such a big population, such a vast territory and such rich re-sources, and what is more, you are said to be building socialism, which is supposed to be superior; if after working at it for 50 to 60 years you are still unable to overtake the United States, what a sorry figure you will cut! You should be read off the face of the earth. Therefore, to overtake the United States is not only possible, but absolutely necessary and obligatory. If we don't, we the Chinese nation will be letting the nations of the world down and we will not be making much of a contribution to mankind.[2]

On February 27, 1957, Mao Zedong added:

> We had no experience of revolution when we first started, and it was only after we had taken a number of tumbles and acquired experience that we won nation-wide victory. What we must demand of ourselves now is to cut down the time needed for gaining experience of economic construction to a shorter period than it took us to gain experience of revolution, and not to pay as high price for it. Some prices we will have to pay, but we hope it will not be as high as that paid during the period of revolution.[3]

On January 13, 1975, Zhou Enlai said:

> On Chairman Mao's instructions, it was suggested in the report on the work of the government to the Third National People's Congress January 1965 that we might envisage the development of our national economy in two stages beginning from the Third Five-Year Plan (1966–1970): The first stage is to build an independent and relatively comprehensive industrial and economic system in 15 years, that is before 1980; the second stage is to accomplish the comprehensive moderniza-tion of agriculture, industry, national defence and science and technology before the end of the century, so that our national economy will be advancing in the front ranks of the world.[4]

In early 1978 this ambition was translated into specific, mid-term programmes, and at the Fifth National People's Congress in March 1978 Premier Hua Guofeng announced the construction of 120 large-scale industrial projects. These include 10 iron and steel complexes, 9 non-ferrous metal complexes, 8 coal-mines, 10 oil and gas fields, 30 power stations, 6 new trunk railways, and 5 key harbours, which will be built over the next 7 years — that is, before the end of 1985 when the Sixth Five-Year Plan is supposed to be completed.[5] Among the expected results is that steel production will reach 60 million tons in 1985, later on reduced to 45 million. The value of industrial output will increase

---

[2] State Planning Commission's Report on Socialist Construction, New China News Agency (in English), Sept. 12, 1977; BBC FE/5613/B11/10.
[3] *On the Correct Handling of Contradictions Among the People,* Beijing, 1966, revised translation, Foreign Languages Press, p. 52.
[4] Report on the work of the government, delivered by Zhou Enlai on Jan. 13, 1975, at the First Session of the Fourth National People's Congress of the People's Republic of China (*Beijing Review,* 1975, No. 4).
[5] Chinese premier announces building of 120 large-scale industrial projects (Beijing, Mar. 6), Xinhua News Agency (Stockholm), 1978, No. 58.

PLATE 1  Steel is obviously a key industrial sector and China is
expected to increase production from roughly 28 million in 1977
to 60 million tons annually in 1985 – now reduced to 45 million
tons per year. The expansion programme will include a steel
combine in Shanghai with an eventual capacity of 6 million tons
per year to be constructed with Japanese assistance. The picture
shows one of the men in charge of the blast furnaces at the
Wuhan Iron and Steelworks – one of the major steel centres
besides Shanghai and Anshan in the north-east.

by an annual rate of over 10% against 4% in agriculture. Nonetheless, China would in
1985 produce 400 million tons of grain against 290 million tons in 1978.[6]

## Foreign Trade

This is, of course, the tip of an iceberg, and the ambitious modernization of the country
will, of course, require a number of technological resources including the training of
engineers, postgraduate students and scientists, at home and abroad, as well as new
technology and technological imports on a substantial scale.

That foreign trade is now being viewed very differently is obvious from an article in
the *People's Daily* pointing out that China's import and export trade accounted only for
less than 1% of the world total in 1977. The commentator says that "this is incompa-

---

[6] Hua Guofeng's report to the NPC FE/5757/i.

tible with China's international prestige and signifies that we have lagged far behind in carrying out the general task of the new period".[7]

That article, which discusses a number of fundamental problems including what is termed "ideological obstructions" to foreign trade, is an outgrowth of the National Conference on Finance and Trade which was convened in Beijing during the summer of 1978. One of the speakers, Vice-premier Li Xiannian, said, after mentioning the need for an increased exchange between agriculture — where "peasants are by and large still doing the work by hand" — and industry, that China should also expand foreign trade. Yet another comment by Li Xiannian has great significance. He said that China should also be exporting more in order to import more advanced technology and equipment to speed the pace of socialist construction. Following on that theme, Yu Qiuli, another of the vice-premiers, also mentioned that new technology should be introduced in a planned way. Furthermore, he went on to say that China "should make foreign trade a channel for making adjustments in the home market so that it is improved. Light industry products and goods needed by the market should be imported."

### Self-reliance and Technology Transfer

Such plans show how the Chinese leadership views self-reliance and self-sufficiency (autarky). In order to understand this situation more clearly, we have to make a clear distinction between self-reliance and self-sufficiency. The latter has two standard interpretations. First, in an economic sense it refers to maintaining oneself without external aid or support. Second, it suggests an attitude of extreme confidence in one's own abilities or worth.[8] Self-reliance, on the other hand, has a much more limited meaning of relying on one's own capabilities which may naturally include intensive trade relations with the outside world. There can be no doubt that in order to reach her ambitious goals, China will use extensively foreign technology and consequently develop both the import and export side of her economy and will not attempt to achieve self-sufficiency.

So we are likely to see China engaged in a massive technology transfer programme — similar to the collaboration with the USSR in the 1950s but this time mainly with the industrialized market economies. The role of technology in the development process is not a simple, easily isolated, and identified phenomenon, but it is generally agreed that technological change is the most crucial single variable in economic growth. The specific role of technology transfer between countries has received serious attention only since World War II, and a better understanding of the transfer process and its inherent obstacles is important in formulating policies designed for faster and easier transfer on a large scale. The removal of obstacles to technology transfer — economic, political, social, legal, and institutional — helps to stimulate economic growth, but a number of inhibiting conflicts remain in the relations between technology suppliers and recipients.

It might be argued that it would not be necessary for China to go outside her borders for such technological, scientific, and industrial development. This may be correct on the assumption that the outside world did not continue to change and develop technologies.

---

[7] The need to "emancipate the mind" in developing foreign trade (NCNA, July 8, 1978); BBC FE/5864/B11/1.
[8] I am indebted to William G. Snead for the use of his definitions in "Self-reliance, internal trade, and China's economic structure", *China Quarterly*, No. 62, (June 1975) 322–8.

It would then have been possible for China to develop gradually using only her own resources.

However, there are a number of constraining factors. One is the obvious relation between the development of science and technology and the development of advanced weapon systems which is likely to remain throughout this century unless there is a major break-through in global disarmament. For this it is still difficult to find any indications, and almost all technologies utilized in modern weaponry are developing very rapidly — electronics and new materials being the most significant areas — and require very considerable resources in order for a country like China to avoid being subjugated to the political pressure of the two superpowers. But the Chinese must consider the outside as a changing world not only in military technologies but also in all technologies, industrially or otherwise. The advanced countries have a leading edge in most sectors, with consequences for surveying and exploiting natural resources, among other things.

For China this problem was further compounded by the mismanagement of the educational system where the training of engineering and scientific manpower was sliding backward in the late sixties and early seventies due to the influence of the so-called "Gang of Four".[9] Consequently, China did not have the resources needed to carry out the ambitious modernization programme outlined by Zhou Enlai at the Fourth National People's Congress in 1975. The logical alternative was to reorganize the economy, including education, research, and industry, and to develop foreign trade. A first attempt to deal with the modernization issues confronting China was carried out in the early 1970s, but the problems were apparently not seriously tackled until a few years later by Deng Xiaoping. His views were presented in three documents related to science and technology, industrial development, and party work, for which he was seriously attacked and criticized in late 1975 and early 1976.[10]

Another alternative would have been to scale down considerably the ambitions of the modernization programme. The opponents — e.g. the Gang of Four and their associates — possibly saw yet a third alternative and were "searching for a formula that would justify the continuation of the radical reforms of the Cultural Revolution while also being responsive to common aspirations for modernization".[11] One of the cornerstones for their development of science and technology concerned the historical interpretation of science with strong emphasis on the contribution of labouring masses over individual genius and the overriding importance of class struggle. From this appears to have followed a neglect of advanced training and research of theoretical and long-term nature, as well as a strong distrust of extended foreign contacts.

## Organizational Models

China has on previous occasions experienced acute struggles over how to organize

[9] This refers to a core group of Mao Zedong's widow, Yao Wenyuan, Zhang Chunqiao, and Wang Hongwen and their associates and followers who, after Mao Zedong's death, were accused of attempting to usurp state power and continue untenable policies.

[10] "Some problems in speeding up industrial development" (draft for discussion, Sept. 2, 1975), *Issues and Studies,* Vol. 13, No. 7 (July 1977). "On the general program for all work of the (whole) party and the (whole) country", *Issues and Studies,* Vol. 13, No. 8 (Aug. 1977). "Several questions concerning the work of science and technology" (also entitled "Outline of briefing on the work of the Academy of Sciences"), *Issues and Studies,* Vol. 13, No. 9 (Sept. 1977).

[11] Prof. R. Suttmeier: "Recent developments in the politics of Chinese science", *Asian Survey,* Vol. 17, No. 4 (Apr. 1977), p. 377.

science and technology. The model which we know from industrialized countries and also from the Soviet Union may be called the professional, or "bureaucratic", model. The other is the Chinese mobilization model, which we know from the Great Leap Forward period in the late fifties and its revival during the Cultural Revolution, whereby very large numbers of people are expected to make their contributions for advancing the technological front.[12] It could be argued that it would not be necessary for China to make a choice between these models and that the solution for the time being lies in a combination of them. It would certainly be incorrect to say that the mobilization approach in science and technology has now completely disappeared in China of today. A mass scientific network in agriculture and the use of amateur scientists in meteorology and seismology are certainly indications of the important role that mobilization still plays. However, this should in no way conceal the fact that the main thrust in all technological sectors lies in the professional approach stressing a high degree of specialization with which we are already familiar in the industrialized countries. The modernization of agriculture and rural areas will also evolve from modern industry and from modern scientific research if present policies are pursued.

The objections against the professional model lie in two different directions. First, it has been seen as corrupting the socialist development within the country. Second, it was also seen as opening the doors to foreign influence, directly as well as through the exchange of goods. From this followed — from the Gang of Four — a commitment to narrowly defined self-reliance or even self-sufficiency in order to counter such influence.

### Goals and Resources

Why do we see changes in China's industry, education, science, and technologies at this particular time? One of the keys to all changes can be found in Zhou Enlai's speech to the Fourth National People's Congress in January 1975, which outlined China's ambitious programme to achieve parity with the advanced industrialized countries by the year 2000. This in turn demanded resources such as educated manpower — engineers, scientists, etc. — which were not available on the required scale. We have reason to believe that the Gang of Four showed serious shortcomings in their understanding of changes outside China and the availability and quality of resources necessary for development inside China.

As long as the objective of rapidly modernizing China was not clearly spelled out — which did not happen until Zhou Enlai's speech in 1975, the conflict was still simmering underneath. There are two conditions which are likely to have acted as an eye-opener to many of the Chinese leaders. One is the increased technology gaps in a number of industrial sectors. The other, obviously related to the first, is the rapid change of military technologies of the industrialized countries which left China far behind. Modern weaponry is characterized by rapidly changing technologies where the research front is moving quickly. With limited research and development resources and limited foreign exchange, China became more underdeveloped — in relative terms — when comparing her technology with other countries. The Soviet Union now accuses China of militarization when China is looking for military technology in Western Europe or in the United States.

---

[12] This and other related issues have excellently been dealt with in a volume by Prof. R. Suttmeier, *Research and Revolution*, Lexington Books, 1974.

However, it would be more correct to see this as an urgent measure taken so as not to be overtaken by development outside China.

Accepting the modernization goals and facing the resource situation in China of 1976, the country had little choice but to change her institutions and develop closer economic, industrial, and commercial ties with the rest of the world.

China has now set as her goal the achievement of full modernization before the end of the century so that the national economy will be advancing in the front ranks of the world. This should in no way be interpreted to mean that China is going to develop science and technology for its own sake. The development goals of the country are clearly deciding the priorities and allocations within the science and technology sector, even if many foreign observers may initially have made a different interpretation. Science and techno-logy must meet different needs in the Chinese society, and we have been given many indications that advanced science and technology may in the past have been temporarily neglected and the intellectuals serving the sector abused — a situation which is now being rapidly remedied.

It must be emphasized that we have to see the changes affecting science and technology in China in a different light than similar changes in advanced industrialized countries. In the already developed market economies most people see science and technology as an instrument which complies with the already existing structures in the society. Some countries, e.g. France, have attempted to build up a structure of complementarity and corrective devices to achieve more or less clearly defined industrial objectives, while the United States provides flexibility to its science and technology through a system of contracts which direct the various research facilities towards national objectives.

## Science and Democracy

The decision on policies relating to science and technology already taken by the Chinese leadership may have far-reaching consequences which may not have been fully contemplated at the time. Now we realize that modernization for a big developing country is no easy undertaking, and there are likely to be some setbacks. However, there can be no doubt that the development of technology and science in China, the more favourable treatment of scientists and sciences, and our increasing contacts provide an opportunity of exchange for mutual benefit.

This liberalization may be essential for the development of a technological and scienti-fic capability. In a review on a recent book, the *Technological Level of Soviet Industry,* Christopher Freeman of the Science Policy Research Unit in England comments on the development of technology in the Soviet planned economy.[13] He says that Japan and West Germany have been much more successful in narrowing the technology gap with the United States than the USSR. He also points out that even if very considerable resources go to military research and production, "the overall performance of the Soviet economy is not such as to justify some sudden increase in NATO military expenditures". The Chinese leadership has possibly for good reasons — not discussed here — a different perception of the Soviet threat. However, more interesting are Freeman's concluding

---

[13] C. Freeman, "Soviet technology", *Nature,* Vol. 272, Apr. 27, 1978 (review of the *Technological Level of Soviet Industry,* ed. by R. Amann, J. M. Cooper, and R.-W. Davies, pp. 575, Yale University Press, New Haven, Connecticut, and London, 1977).

remarks on the association between democracy and the progress of science and technology. He sees this association as a complex one and of a long-term nature. But he says with reference to the Soviet study that it "lends overwhelming support for the view of those communists who have been arguing that the principal constraint on the progress of the socialist economies is the lack of initiative, participation, and independent criticism at all levels of economic management, politics and science".

The changes in technology and policy in China have not been discussed in exactly the same terms. There can, however, be no doubt that the new situation has created a very different situation where engineers, scientists, and other researchers can discuss and act much more freely than in the past 10 or more years. Related to this is the readiness to enter in discussion with foreign scientists and use foreign technology to China's advantage. Equally important are the decisions of the leadership to allocate resources after having had thorough consultations with the scientific community. In commemorating May Fourth, the *People's Daily* in spring 1978 said that the scientific spirit and the democratic spirit are inseparable in the struggle to build a modern and powerful socialist country. Science and democracy were the two banners of the May Fourth movement that swept China in 1919.[14]

**The Future**

Finally, technology transfer during the last quarter of the twentieth century is likely to differ from that of earlier periods. There are many reasons. First, the demand for technology is increasing as countries and regions are drawn into much closer interaction and as differences in the standard of living among nations are increasingly obvious. Second, technology is much more intimately related to science than in the past. Thus, the developing countries must attempt not only to approach, by adaptation, a much higher level of science and technology, but they must also reach for science and technology that are continually and rapidly moving to higher levels. Practically all of this advancement takes place in the industrialized countries.

An attempt will be made in the following chapters to relate the changes to the existing situation and the needs in China. Here we have to remember that China is a country with a still dominant majority of her manpower in agriculture. We are also looking at a continental country with manpower resources far exceeding any other country. At the same time China has set up structures in science and technology which, on the one hand, should facilitate her development strategy and, on the other hand, should be one of the instruments in moving towards socialism and — Mao's conception — creating a new man. Consequently, it has been considered necessary to make a fairly detailed presentation of policy issues and the science policy debate in China.

Finally, it must also be emphasized that the situation in China is rapidly changing. It may still be too early to say that the country is beyond the age of need and has reached the age of choice. The structure for encouraging innovations and change through technology and science have not been fully worked out. There may still be some uncertainty about the balance between centrally planned solutions and local, lower-level contributions, and the role of foreign technology is likely to be a changing one over the years to come.

[14] *People's Daily*, Commentary on youth, science, and democracy (June 5, 1978); BBC FE/5808/B11/6.

Consequently, the debate on science and technology as reflected in the news media over the past couple of years will frequently be used, hopefully, to shed some light on the Chinese view of science and technology development.

# 1. Today's Situation

### Pace and Direction of Technological Change

The availability of modern, transferable technology has increased the interdependence of countries and shifted developmental concern from indigenous technological evolution to the transfer of technology from one country to another. The concern about the technological gap between developing and industrialized countries is of relatively recent origin. The economic progress of the industrialized countries has been achieved only in the last century. Uneven development — which is at the core of technology transfers — has been the rule as nations move through various modes of production. India was economically superior to Britain before becoming an exploited colony, while Britain, once the centre of industrialization, is rapidly declining in relative importance. India is now slowly reemerging and gathering strength, although more slowly than another potential giant, the People's Republic of China. Many of the national superiorities of the 1970s are likely to be transitory, and technology transfer will play an important role in future transitions. However, the ability to control the technology transfer, rather than the availability of technology, is going to be the decisive factor.

The American analyst, Goulet, concludes that "technology is perhaps the most vital arena where cultures and subcultures will either survive or be crushed. Their absorptive capacity will be tested in this arena."[1] Borrowed technology contributes to the expanding vitality of the borrowing culture only in so far as it both nourishes and draws nourishment from the activities that led to borrowing in the first place. If it becomes a substitute for native forces, it is destructive rather than constructive.

For some years it has been fashionable to discuss appropriate and intermediate technologies and their relevance in developing countries.[2] Much of this has been a discussion in isolation — often of a highly theoretical nature. In the meantime the industrialized countries have continued to develop *their* technologies in manufacturing, agriculture, services, etc. Suddenly the appropriate technology movement is receiving ample funds and attention — from the World Bank, the US Agency for International Development, and so on. In the meantime the justification for appropriate technology *may* have undergone a considerable change. Because of the pace and direction of technological change in the industrialized countries the justification — or rather the possibilities — for intermediate technologies may have been seriously eroded. From this it follows that the formal system for science and technology — deservingly criticized for poor performance in many countries — should possibly be seen in a different light.

The very considerable change in technology and science policy in China appears to be

---

[1] D. Goulet, "The paradox of technology transfer," *Bulletin of the Atomic Scientists,* Vol. 31 (June 1975), pp. 39–46.

[2] The comments in the following paragraphs were first published as a short note in the *Lund Letter on Science, Technology and Basic Human Needs,* a newsletter published by the Research Policy Programme at the University of Lund, Sweden.

an example of such a realization. The new emphasis is naturally an indication of an earlier unsettled political situation which made proper policy decisions very difficult. However, there is much more to the heavy emphasis on the modern large-scale technologies and use of imported technologies. We see this as a reflection of a realization that the technological development of the industrialized world is deciding not only the direction but also the pace of technological change, and that China has decided to take up the challenge. In a number of statements Chinese policy-makers have been pointing out the existence of serious technology gaps and how to deal with them.

### Technology Gaps

In 1977 the noted Chinese scientist Qian Xuesen discussed the gap between China's science and technology and the advanced countries of the world.[3] In his views and comments, which are personal, he deals with four questions related to the gap. The first is whether or not there exists a disparity between the level in China's science and technology and advanced-world levels, and he says that there are some things in China which come close to or which surpass advanced-world levels. But they represent only a part of the whole, and a relatively small part at that, and in most cases China is relatively backward. Furthermore, he says, among those scientific and technological undertakings in which China has achieved relatively better successes and has surpassed foreign countries in terms of overall results, the technical level of some of the machinery and equipment is not high.

He then goes on to discuss the second question — whether China should gradually narrow the disparity and catch up with and surpass the advanced-world levels. The speed and direction have been critical issues, and Qian Xuesen's views are not always very illuminating. In essence he says that socialist economic construction and the national defence requires the narrowing of the disparity.

On the third question, e.g. whether China is able to catch up with and surpass the industrialized world, he is equally vague by referring to the superiority of the country's socialist system. He points to the contradiction between the socialization of science and technology and private ownership under the capitalist system — a contradiction which cannot be resolved. Consequently, according to Qian's views, this has placed an obstruction in the path of the development of science and technology in market economies. If we look at the recent developments in science and technology in advanced industrialized countries, this can only be partially correct. Nonetheless, Qian answers the third question in the affirmative and says that, in the final analysis, the interests of the individual, the collective, and the State are in accordance with one another in China. However, if we accept at least some of the criticism levelled at the political opponents known as the Gang of Four, this harmony cannot have been prevalent in China in recent years.

In his reply to the final question, Qian Xuesen discusses how to bring the superiority of the socialist system into full play so as to enable China's science and technology to catch up and surpass the advanced-world levels. Even here the article lacks specificity, but the substance can be seen in the many reforms and measures which have been announced and implemented since the article was published.

---

[3] Qian Xuesen, "Science and technology must catch up with and surpass advanced world levels before the end of the century", *Red Flag*, 1977; BBC FE/5563/B11/6.

PLATE 2. The textile industry employs 2.8 million staff and workers. Artificial fibres from the domestic petrochemical industry are increasingly used in order to relieve the pressure on agriculture for industrial raw materials. The sector has a number of smaller, local, enterprises, and the picture shows a commune-run silk filature in Linxian County in Henan Province.

### Reorganization of Science and Technology

The somewhat gloomy views of Qian Xuesan on the present situation are being reiterated in a statement by Fang Yi, Vice-president of the Chinese Academy of Sciences, who, at the end of December 1977, said that "China's science and education are in such a state that virtually everything needs to be done". In order to deal with the situation, the Party Central Committee and the State Council have made a number of major decisions and have taken the following effective measures.[4]

1. A state commission for science and technology has been set up. Its responsibilities will be the overall planning, co-ordination, organization, and administration of the country's scientific and technological work.

2. A system of directors assuming responsibility in research institutes under the leadership of the party committees has been approved. The leadership of many scientific research institutions and a number of universities and colleges has been reorganized and strengthened.

3. A programme for the development of science and technology and that of education is being drafted.

4. The administrative system has been changed in order to make full use of local and

---

[4] Comrade Fang Yi on heartening progress in China's science and education, Xinhua News Agency (Stockholm), 1977, No. 314 (Dec. 31). Fang Yi was making a report on the situation in China's science and technology on Dec. 27, 1977, at the seventh session of the standing committee of the fourth national committee of the Chinese people's political consultative conference.

central initiatives. Some of the scientific research institutions, which were transferred or abolished, have been restored and others are under consideration.

5. Science conferences or teachers' meetings have been held in a number of places to award those with outstanding achievements and to exchange experience. The restoration of titles for technical personnel has been carried out in order to encourage improvement in professional skill, assess technical proficiency, and strengthen the system of specific responsibilities.

6. Academic and working conferences have been held to implement the principle of "letting a hundred flowers blossom and a hundred schools contend". The national scientific and technological association has been revived and various scientific societies have resumed their work.

7. A new system for enrolling students in universities and colleges has been implemented, and large numbers of promising students have come forward.

8. The compilation of a new set of standard textbooks for the whole country is being carried out under the responsibility of the Ministry of Education. The principle followed is to condense the teaching material and do everything in order to provide the young people with the latest scientific and technical knowledge.

9. While keeping to the principle of independence and self-reliance, China will strive to learn advanced science and technology from foreign countries and promote international academic exchanges.

10. Scientific and technical personnel will be guaranteed at least five-sixths of the week for professional work.

11. Science and education will have their funds coming from the state budget appropriately increased in order to accelerate and expand development.

12. Science and education will receive more publicity and more efforts will be made to spread scientific knowledge.

## The National Science Conference

In September 1977 the State Council announced a National Science Conference to be held in early 1978. A planning conference was held in October and subsequently a draft plan for science and technology was prepared covering the country's needs and tasks of the coming 8 years up till the end of 1985. Penetrating discussions and consultations on the plan were carried out in December 1977 and January 1978, and more than 20,000 people were involved. The regional units of the People's Liberation Army and provincial science and technology bodies were meeting all over the country, obviously to discuss matters of relevance for the national plan. A plan was then presented at the National Science Conference which was held in late March and early April 1978, and a new series of provincial follow-up meetings took place all over the country in the succeeding months.

A growing awareness and eventually an in-depth knowledge of the problems confronting China in her modernization programme is likely to have developed among planners, technicians, and politicians in a number of national conferences which met in late 1977 and early 1978 and have continued with an accelerating pace since the National Science Conference. A selected number of these conferences are listed in Table 1, which shows the range of topics discussed. The increasing emphasis on improving the educational system, engaging in technology imports, and collaborating with scientists abroad should be seen, at least partly, as an outcome of the national conferences. The

TABLE 1. *National Conferences in 1977 and 1978*

| Topic | Period | Location |
|---|---|---|
| Post and tele-communica-<br>tions | 1977 July 00–30 | Beijing (?) |
| Electronics industry | 1977 Nov. 7–Dec. 4 | Beijing |
| Power | 1977 Nov.– Dec. | Beijing |
| State farms | 1978 Jan. 00–25 | Beijing (?) |
| Farm mechanization | 1978 Jan. 4–26 | Beijing (?) |
| Building materials | 1978 Jan. | Beijing (?) |
| Technical innovations | 1978 Jan. 15–22 | Yentai |
| Profits | 1978 Jan. | ? |
| Metallurgy | 1978 Jan. 00–16 | Beijing |
| Coal | 1978 Jan. ?–Feb. ? | ? |
| Civil aviation | 1978 Feb. | ? |
| Science | 1978 Mar. 18–31 | Beijing |
| Railway | 1978 Apr. 15–23 | Beijing |
| Capital construction | 1978 Apr. | Beijing |
| Education | 1978 Apr. 22–May 16 | Beijing (?) |
| Transport/<br>communications | 1978 May 2–11 (24?) | Taqing |
| Textile industry | 1978 May ?–23 | Beijing (?) |
| Oil and chemical<br>industries | 1978 May 23–June 11 | Taqing |
| Light industry | 1978 June 2–10 | Beijing |
| Medical science | 1978          –12 | Beijing |
| Construction machinery | 1978 June | Wenchiang |
| National supplies | 1978 June 19–July 4 | Taqing/Beijing |
| Finance and trade | 1978 June 20–July 9 | Beijing |
| Capital construction in<br>agriculture | 1978 July | Beijing |

Sources: BBC *Summary of World Broadcasts,* Part 3, *The Far East,* and Xinhua News Agency (Stockholm).

Eight-year Plan, 1978–85, gives prominence to eight comprehensive areas of science and technology, namely energy resources, materials, electronic computers, lasers, space science, high-energy physics, genetics, and – naturally – agriculture will cover both basic and technical sciences. Included in this approach are 108 items which have been chosen as key projects although these have not yet been made public. When the plan is fulfilled in 1985, according to Fang Yi China is expected to approach or reach the advanced-world levels of the 1970s in a number of important branches of science and technology with a resulting reduction in the gap to about 10 years. [5],[6]

Similarly, medium-term plans for 5 and 8 years have also been drawn up for the Academy of Social Sciences. This Academy has been created as a new entity out of the department of social sciences which until recently was part of the Academy of Sciences.[7]

Organizational measures are introduced which will help China to develop greatly increased science and technology capability. A key element is the system of "individual responsibility for technical work" in the scientific research institutes and "the system of

---

[5] Vice-president of Chinese Academy of Sciences on policies guiding work, Xinhua News Agency (Beijing, Mar. 30, 1978), No. 10668.

[6] Fang Yi's report at the National Science Conference, Xinhua News Agency (Beijing, Mar. 28, 1978), No. 10666.

[7] President of Chinese Academy of Sciences Kuo Moruo calls for new vitality in philosophy and social sciences, Xinhua News Agency (Stockholm), 1978, No. 63.

division of responsibilities among institute directors under the leadership of party committees". On judging the work of the party committees of the scientific research institutes, Deng Xiaoping stressed that "the main criterion for judging the work . . . should be the successful fulfilment of the task of producing as many scientific results and training as many highly competent scientific and technical personnel as possible".[8] It is interesting to note that personnel are to be trained for management of scientific research, apparently at the Chinese University of Science and Technology, which is under the leadership of the Academy of Sciences as well as at the new University of Science and Technology in Harbin.[9]

## Technical and Scientific Collaboration

Related to the goal of catching up, Vice-premier Deng Xiaoping in his opening statement at the conference stressed that: "it is not just today, when we are scientifically and technically backward, that we need to learn from other countries — after we catch up with advanced-world levels in science and technology, we shall still have to learn from the strong points of others".[10]

The present interest in collaboration is reflected in comments made by Li Chang at the National Science Conference. He said that the postgraduate school of the University of Science and Technology and other research institutes will invite outstanding foreign scientists to conduct seminars. Invitations from prominent foreign and friendly research institutions will also be accepted so that high-level Chinese scientists and technicians can take part in their research projects.

More far-reaching suggestions for scientific contacts with the industrialized West were revealed in late summer 1978. The Chinese started negotiations with a number of countries to find out if they were willing to receive Chinese students and scientists in large numbers for extended training of around 2 years. It was suggested that Sweden might initially receive 100–200 and the United Kingdom up to 1000 and Canada 500.[11] Judging from these tentative figures one might envisage a situation where more than 10,000 Chinese students and scientists were receiving postgraduate and specialized training in the industrialized market economies. Whether a programme of this magnitude will materialize is uncertain at the time of writing. However, the Chinese interest in a training programme of this kind has certainly indicated that the need for specialized training in natural sciences and engineering exists and cannot easily be met within the country.

The apparent interest in having scientists trained abroad should be related to a categorization made by Deng Xiaoping in 1975.[12] In his "Outline Report" he divides the scientists into four groups: (1) those trained before the liberation, including those trained abroad; (2) those sent abroad for training after liberation, (3) those trained in China after

---

[8] Vice-chairman Deng Xiaoping on party leadership over scientific and technological work, Xinhua News Agency (Stockholm), 1978, No. 71.

[9] Vice-president (Li Chang) of Chinese Academy of Sciences on policies guiding work, Xinhua News Agency (Beijing, Mar. 30, 1978), No. 10668.

[10] Vice-chairman Deng Xiaoping on self-reliance and learning from others.

[11] Private communication to the author and the *International Herald Tribune*, Aug. 4, 1978 (China to send students).

[12] R. P. Suttmeier refers to the categories in his article "Recent development in the politics of Chinese science", *Asian Survey*, Vol. 17, No. 4 (Apr. 1977), pp. 375–92.

liberation; and (4) those selected from among workers, peasants, and soldiers for training. The last two categories are seen as being of limited use in China's scientific and techno- logical development, while the others are to be "boldly made use of". The first group, in particular, is said to have broader knowledge and more experience, while the second will provide the "mainstay in work".

## Publications

Information processing is another key element for development. The publication of scientific and technical journals has apparently been fully resumed. The library at Qinghua University in Beijing, for example, stocks close to 2000 Chinese-language periodicals published in China and all related to science and technology.[13] Many of these are still for internal circulation only and sometimes may only be available on an exchange basis between research institutes, universities, colleges, etc. That only a limited number of copies are available is obviously the case for those which are using pasted photographs for illustrations. The periodicals cover a large number of technical fields with prominence given to electronics and metallurgy. A large number of them are abstract journals on foreign literature, while others contain full translations of foreign articles — all of which is similar to the situation in periodicals publication before the Cultural Revolution. A small number of periodicals are available to foreigners. Appendix V lists most of the periodicals in science and technology which were available in early autumn 1978.

[13] Information from the author's visit to China in late March and early April 1978.

# 2. An Historical Perspective

CHINA'S attempts to modernize science and technology must be seen in their proper historical perspective. According to a common belief there was never any science or technology in China at all. It may seem strange that this should ever have been believed. Yet this was the impression of sinologists when Joseph Needham began his investigations.[1] And it was repeated by generations of superficial observers who knew nothing of the Chinese historical literature on science and technology.

It is now evident that between the fifth century BC and the fifteenth century AD, Chinese bureaucratic feudalism was much more effective in the useful application of natural knowledge than the slave-owning tradition of classical cultures or the serf-based military aristocratic feudal system in Europe.

If the scholar—gentry systematically suppressed the occasional development of mercantile capital, it was not in their interest to suppress innovations which might be put to use in improving the production of the counties or provinces in their charge. Needham argues that even if China had an apparently limitless reservoir of labour power, it remains a fact that we have not so far met with a single case of the rejection of an invention due to the explicit fear of unemployment.

But what about the notion of "stagnant" China, a country that has obviously been the donor of so many discoveries and innovations which have acted like time-bombs in the social structure of the West? This concept of stagnation, born out of Western misunderstanding, was never truly applicable. What happened was that China's slow and steady progress was overtaken by the rapid growth of modern science in Europe after the Renaissance.

During the first fourteen centuries of the Christian era, China transmitted to Europe an abundance of discoveries and inventions which were often received in the West with no clear idea of where they had originated. The technical innovations travelled faster and further than the scientific thought. By the time of the eighteenth century we reach the beginning of the modern period when science has become a worldwide enterprise in which China participates alongside all other cultures.

With further reference to Needham, he and others have demonstrated that a number of popularly-held antitheses are historically false. He mentions among other things:

(1) The Chinese made sure that their tombs faced due south but Columbus discovered America.
(2) The Chinese planned the steam-engine's anatomy but Watt applied steam to the piston.
(3) The Chinese used the rotary fan but only for cooling palaces.
(4) The Chinese understood selection but confined it to the breeding of fancy goldfish.

[1] These comments are to a considerable extent based on "Science and China's influence on the world" by Joseph Needham, pp. 234–308, in the *Legacy of China* (ed. by Raymond Dawson), Oxford University Press, 1971 (1964), paperback.

However, the truth is that the inventions and discoveries of the Chinese were mostly put to great and widespread use, but under the control of a society which had relatively stable standards.

There can be no doubt that in the opening phase of modern science, when mechanics, dynamics, and celestial and terrestrial physics came into being in their modern form, the Greek contribution had the greatest share.

In technological influences before and during the Renaissance, China occupied a dominant position. Her contributions included, in particular, iron and steel technology, the invention of gunpowder and paper, the mechanical clock, and basic engineering devices such as the driving belt, the chain-drive, and the standard method of converting rotary to rectilinear motion together with segmental arch bridges and nautical techniques such as the stern-post rudder.

Why has there been this difference between China and the West in the recent application of science and technology to economic and social change? As for China, social analysis will assuredly point to the nature of agriculture, the early necessity of massive hydraulic engineering works, the centralization of government, etc. Thus it was radically different from the patterns of the West.

What was then the instability of Europe due to? Needham and others prefer to think in terms of the geography of what was in effect an archipelago, the tradition of independent city-states based on maritime commerce and jostling military aristocrats ruling small areas of land, the exceptional poverty of Europe in precious metals, the continual desire of Western people for commodities which they themselves could not produce, and the inherently divisive tendencies of the alphabetic script.

It may be that Europe had some of the secret of the specific creativeness for which the time had become ripe. China eventually experienced the necessity for entering the world community which these great forces were forming, and that is where we are now.

The People's Republic of China in 1949 inherited a system of scientific and educational institutions which had been modelled on their equivalents in Europe and the United States. Some useful research was performed in these institutions and a number of eminent scientists received their initial training in them. However, these institutions had little or negligible impact on the social and economic development of the country in the period up to 1949. After that the new government reorganized science and educational institutions along the lines existing in the Soviet Union.

It was envisaged that the new orientation would achieve two objectives: first, to expand China's scientific and technological capabilities as quickly as possible; second, to orient scientific research to the future technological needs of China's modern industry which was to be initially constructed with technologies imported from the Soviet Union. The new system for science, technology, and education was in fact appropriate to this strategy of industrialization. However, once the objectives changed, i.e. in order to achieve a more balanced development of rural areas and agriculture where China's manpower resources were given more consideration, the existing system was no longer relevant. Organizational reforms were undertaken in the late 1950s and the early 1960s, but the most drastic experimentation with scientific and educational institutions took place during the Cultural Revolution.

In 1974 a survey[2] was carried out at the Research Policy Programme in Sweden when

[2] B. Berner, *China's Science Through the Visitors' Eyes,* Research Policy Programme, University of Lund, Mar. 1975.

several hundreds of scientists, mainly from the United States and countries in Western Europe, who visited China in the early 1970s were interviewed through a questionnaire. They were, among other things, asked the following:

Do you think that the Chinese way of approaching science and technology development after the Cultural Revolution can serve as (a) a model for developing countries, or (b) a model for developed countries?

The survey revealed that 65% of the natural scientists and 57% of the medical scientists thought that the Chinese way was suitable for developing countries. One set of comments stresses the success of the means with which China has begun to tackle some of the particular and pressing needs of developing countries. This relates to such problems as agricultural development, health care, illiteracy, urbanization, employment, ecology, and environment. The orientation of science and technology to these immediate and enormous problems is seen as relevant for other developing countries.

However, many visiting scientists — even if they agree in the comments referred to — still very much doubt the relevance of the Chinese approach in science and technology for other developing countries. They argue that there is no possibility of instituting something like the Chinese science and technology system until a socialist revolution has changed the national and international power structure at present acting upon most developing countries.

The survey also revealed, not surprisingly, that a majority of medical as well as natural scientists do not consider science and technology in China to be relevant for industrialized countries. However, it may be of some interest to look at the positive reactions. It is argued that Chinese science has greater social relevance, stresses self-reliance, deals with problems such as pollution, energy problems, and work organization which also are great problems in the industrialized countries. In addition to this, many are also impressed by the participation of the masses in science and technology and the emphasis given to local problems. And there is — according to these scientists — less bureaucracy, less centralization, and more democracy in the application of technology in China.

Many people have pointed out after visiting China that industrialized countries need to close the considerable gap between elite science and practical application. But it is doubtful if such a policy can be instituted outside China without considerable changes in social organization and values. Finally, it was stressed that the heavy emphasis on application is not relevant for already industrialized countries. Some of the visiting scientists pointed out that the developed countries will in the immediate future bear the burden of the academic research which, in their opinion, will lead to the greatest long-term advances.

It is evident from the survey — as seen by the replies to the two questions — that science and technology in China is judged very differently by different observers — at least when it comes to taking the system outside China. It appears from the report that non-US scientists seem to have been more careful in their judgements than their US colleagues. More interesting, however, is the fact that those favourable to China's science policy system as a model in developing and/or developed countries appear to have a better previous knowledge of China.

As a conclusion to the survey there seemed to be a general agreement among those interviewed that in China:

(1) resources and techniques are behind world level; and
(2) international contacts are much less than in their own country.

Moreover, all agreed in a statement that the social relevance of research as well as the degree of contact between research work and practical application were greater than in their home countries, and that these aspects seem to have received continued emphasis since the Cultural Revolution.

However, I think that many of us may have fallen into the trap of making China into some kind of an ideal, both in our comparisons as well as in seeing China as a country which really has a better ability for handling problems. Looking at the technology and science debate in China over the past couple of years gives a somewhat different picture — with missed opportunities, increasing technological gaps, and ambiguity over the role of professionals and intellectuals. This is, of course, due to many factors — maybe partly a reflection of a Chinese misunderstanding of the speed and direction of technological change outside the country. Another important factor is the unsettled political situation inside the country after the demise and death of Mao Zedong, which may temporarily have made proper policy decisions very difficult. In particular, China appears to have neglected large-scale projects not yielding immediate results but requiring long-term commitment of resources — finance and equipment as well as highly qualified researchers. That situation is now changing.

# 3. China's Objectives

CHINA'S policy for scientific research and technological development is influenced through political, ideological, and other societal values which should be termed long-range objectives. These are international as well as domestic in character. The internal long-range objective may be taken to be the attainment of a perfect communist society characterized among other things by the principle of distribution according to needs. Internationally, the long-range objectives are a full realization of China's ideal of justice and equality throughout the world. Such objectives would in international relations mean in part an elevation of the Third World countries into a central position in the world.

Since the defeats she suffered at the hands of various imperial powers in the middle of the last century, China has engaged in finding the formula which would enable the country to regain what it considers its rightful place among nations. It is only since the Chinese Communist Party gained full control over the country in 1949 that it has been possible to move systematically towards the objectives of modernization. The Chinese leadership very early stressed that the development of technology and science was not only an important tool for economic and industrial development, but also for social development and for creating a scientific outlook among the people.

The concern for industrial modernization and technological development, both in the civilian and military sectors, should not be evaluated only in our frame of reference but also with the historical experience of China in mind. This viewpoint was underlined in a joint editorial by the *People's Daily, Red Flag,* and *Liberation Army Daily* at the end of the Fifth National People's Congress.[1] With a quotation from Mao Zedong we are reminded that from 1840 to 1945 China suffered great humiliations at the hands of imperialist countries. Almost all imperialist countries "whether large, medium or small" committed aggression against China. The editorial states that this was due to two factors: the corrupt social system and the backward economy and technology. In the case of the first factor, China has already put her house in order even if it is admitted that "the solution is still incomplete because class struggle still exists". But for the second factor China has only achieved some degree of change, and the country "will require several more decades to bring about a complete change". But if this is not done within the next few decades "it will be impossible for us to avoid being pushed around again".

We should also remember that China has summarized her political objectives in the concept of reducing the three differences, e.g. those between manual and intellectual workers, between agriculture and industry, and between rural areas and the cities. A commitment to reducing the differences has long-term consequences for her policy on science and technology, the role of intellectuals and education, etc., all of which will be discussed throughout this book.

Most development economists have over the past couple of decades changed their

---

[1] Editorial by the *People's Daily, Red Flag,* and *Liberation Army Daily* entitled "Transform China in the spirit of the foolish old man who removed the mountains" (Mar. 6, 1978); BBC FE/5757/C/1.

views on the road to prosperous development. Today it is realized that a high rate of growth of the national product and a rapid expansion of industry do not provide all the solutions. Programmes to meet the basic needs of the population majority are essential. Furthermore, it has also been understood in many countries that small-scale industry and services — "the informal sector" — as well as the agricultural sector, are going to play an important role — at least in terms of employment for a large number of years.

Those concerned with science, technology, and development have experienced a similar change. At first, science and technology were seen as the key instruments for solving all problems. They were to be the motor of industry which would propel the developing countries to a high level of prosperity. The structure and approach were borrowed from industrialized countries.

In the late 1960s it was realized that science and technology had not achieved what was expected of them. All the borrowing from the industrialized countries, the dependency it led to, and other factors, became evident. The notions of integrated rural development, intermediate/appropriate technology came to be seen as the new possibilities which were going to provide employment, equality, and development.

Some people have attempted to provide a new analytical framework for a different science and technology structure in developing countries. Outside Latin America there is hardly any developing country except China which has attempted to shape a new structure. China has since the Cultural Revolution tried to find new pioneers of knowledge in industry and still more in agriculture. The power to challenge and change the direction of the search for new knowledge is directly related to political power. Consequently, the new directions in scientific and technical development and the capability to decide what potential new knowledge should be searched for will be consolidated only if the political power base remains the same.

There can be no doubt that it is essential for China to create a materially rich society in order to attain the long-term domestic and international objectives. And there can be no doubt, either, that technology will play a very important role in this process. In order to achieve such economic objectives, the Chinese economy is called upon to perform the following tasks:

(1)  meet basic needs for food, shelter, and clothing;
(2)  provide essential consumption goods;
(3)  generate economic surplus that may be used for investment and foreign aid;
(4)  provide institutions which support the qualitative improvement of human and capital resources;
(5)  provide an industrial and technological base for the production of national defence goods;
(6)  maintain and increase economic independence from foreign and natural influences.

In this China's technology policy must fulfil three major objectives: first, to achieve a better utilization of natural, capital, and human resources; second, to raise the country's overall level of technological sophistication; and third, to decrease production costs and improve product quality. Achieving these objectives would ensure a higher standard of living, a greater range of long-term options, and more bargaining power in the world market.

China's technological policy, in the modern sector of the economy, is still focused primarily on the problem of catching up — technologically, industrially, and economically – with the most advanced countries of the world, including Japan. This preoccupation should not be seen as a mere desire to imitate or emulate what industrialized countries have been doing. The objective of China's technology policy should be seen as a conscious attempt to obtain the technological and industrial means for achieving the political position to which it is entitled by virtue of its culture, history, and social traditions. Furthermore, China's technology policy also aims at achieving social objectives, particularly in the traditional sector of the economy as the Chinese leadership is committed to egalitarian goals and a high degree of participation.

Capital and human resources in the traditional sector are wasted if a country's technology policy is geared to meet only the technology demand arising in the modern sector of the economy. This must be obvious as soon as one realizes that the majority of a developing country's human resources is in rural areas, a situation which will continue for a number of years. Further, it has been argued in the past that rapid industrialization based on the modern sector would bring all the fruits of modernization. It has now been generally accepted that economic change must take place in all sectors of the economy in order not to cause overwhelming social and political problems. Consequently, technology policy and its implementation in the traditional sector must be given equal emphasis to that in the modern sector.

Technology policy in China today cannot be fully understood without realizing the underlying dual-economy development strategy. This gives rise to a variety of technology demands which have to be met with distinctly different technology policies. Consequently, there is increasing evidence to show that China is pursuing a "dual policy" in organizing research and the distribution and application of results.

In order to understand this, it is necessary to differentiate among the various sectors where research and development (R & D) results are required. Long-term projects, with great demands on resources, generally require specialized research workers with specialized education. The control of such resources – financial and manpower – also requires substantial organizational resources. These projects are found particularly within the defence and the modern industrial sectors. The priorities are set here by national agencies, and popular participation in the decision-making process is almost non-existent.

The intense political debate raging in the scientific and technological circles in China over the past couple of years has been resolved temporarily. It has been clearly stated that scientific research must precede production, and the outcome has been strongly influenced by an increasingly strong awareness that China's material base must rapidly expand and that science and technology are important and efficient instruments for achieving this.

This is particularly so in the modern industrial sector. Here the improvement in production efficiency is the last stage in a complex sequence of technological change which basically has three phases.[2] First, there is research and development leading to a new idea or invention, usually in the form of a technically feasible prototype. Second, the invention is developed and modified so as to improve it in certain respects, culminating in the first application that is economically feasible. Third, the recognition of the product as a more efficient technology results in its further application within the same enterprise and its introduction to other firms within the industry. The successive stages often overlap

[2] This sequence is taken from the introduction in L. Nabseth and G. F. Ray, *The Diffusion of New Industrial Processes,* Cambridge University Press, 1974.

and the lines of demarcation between them are blurred. However, the shortcomings of China's science and technology policy over the past few years, which are now in the process of being remedied, lie in three different areas. The first stage, research and development, has been neglected in a number of important sectors of the economy as a consequence of insufficient allocation of resources. Second, there has been a lack of coordination and discipline in all three stages. Third, a growing but unmet demand for specialists of different kinds has added to the problem.

As various economic sectors in a society require different types and mixtures of R & D resources, the distribution of resources between basic research and grassroots applications must also vary in the different sectors. The level of development and the consequent need for and availability of new resources has implications for the distribution of resources along the various parts of the innovation chain. This in turn has consequences for the forms utilized to organize R & D.

This balance comes out very clearly in the following five principles which are now guiding the development of technology and science in China.[3] These principles, which reflect a strong desire to balance counteracting forces and needs, indicate a pervading theme in Chinese policy, especially science policy. They relate to the Yin-Yang dichotomy which is exemplified by pairs such as light—dark, female—male, earth—heaven, and have strong historical roots which some trace to the Taoist philosophy.

(1) On the relationship between politics and technique, the report stressed that it is wrong not to criticize the tendency of ignoring politics; on the other hand, it is also wrong not to encourage scientific and technical personnel to study professional knowledge.

(2) On the question of scientific and technical personnel integrating with the workers and peasants, it is wrong not to criticize the tendency of belittling the role of the masses in scientific research, and it is equally wrong not to give full play to the role of specialized research institutes and specialists.

(3) On the relationship between scientific research and production, it is wrong not to call on scientific and technical personnel to do research and solve urgent problems in actual production; on the other hand, it is equally wrong to overlook or deny the importance of research in basic theory and the necessity of laboratory work.

(4) On the question of remoulding intellectuals and enlisting their service, it is wrong to assume that the remoulding of their world outlook has more or less been completed; on the other hand, it is equally wrong to maintain that they cannot be of any use until they have remoulded themselves completely.

(5) On the question of the relationship between Marxist philosophy and the natural sciences, it is wrong to deny the guiding role of philosophy in relation to the natural sciences; on the other hand, it is equally wrong to think that the former can substitute the latter and that concrete conclusions on specific scientific problems can be arrived at by relying on the general principles of philosophy.

Here we should remember that most professionals in advanced countries regard social change as a consequence of technological development. The view prevailing in China during the Cultural Revolution has been that social changes should occur simultaneously or should even precede technological changes. Consequently, the Gang of Four attempted a full integration between production and research in order to reduce or even eliminate any barriers and distinctions between researchers and the people.

This should be seen in the light of the historical experience of the now advanced countries. Here the scientific and technological progress achieved over the past 200 years followed rather than preceded social transformation. This kind of development has its

---

[3] Chung Ko, "The struggle around the outline report on science and technology", *Beijing Review*, No. 44, 1977, pp. 5—8.

parallel — at least partially — in the rural areas of China. The agricultural sector has undergone a tremendous social transformation since 1949 exemplified by collectivization, the commune reform, and large-scale mobilization of people for development projects. It is also in this sector of the economy that the mass scientific approach appears to have been most successfully implemented. But the question of whether social change should precede or follow from research and technological development has been a very fundamental one in China and may not yet have been fully resolved.

The short-term objectives are, internationally, to increase national independence so as to counter threats of the superpowers, and to prevent the Third World, particularly its natural resources, from falling into the hands of the superpowers. Primary importance is then placed on finding workable arrangements which would ensure the national security of China and the continued progress towards the achievement of the long-range objectives. This also leads to an obvious need to rapidly develop Chinese industrial strength.

The development of science and technology in China must be seen against the characteristics of the country. First, China is a developing country with a predominantly rural population majority, most of whom are still engaged in agriculture. Second, China is a very big country with great regional variations and a diverse economy. Third, and not least important, the People's Republic of China is committed to socialist planning with the ultimate aim of developing a communist society.

The influence of such characteristics and a changing perception of what science and technology should accomplish partly comes through by looking at the priority areas in China's two science and technology plans: the recent one for 1978—85 and the earlier one for 1956—67 (Table 2). Today the needs of agriculture emerge as the first priority,

PLATE 3. With roughly 10% of the world's arable land and approximately 22% of the population, China considers farmland construction and increased land productivity as essential. The picture shows desert control workers studying soil amelioration in an oasis in the desert area.

TABLE 2. *Priority areas in China's science and technology*

| 1978–85 | | 1956–67 |
|---|---|---|
| 1. Agriculture | 10. | Chemical fertilizers |
| | 9. | Yellow and Yangtze river control |
| | 11. | Eradication of diseases |
| 2. Energy | 1. | Peaceful atomic energy |
| | 5. | Petroleum and mineral exploration |
| | 7. | Fuel technology |
| 3. Materials | 6. | Metallurgy – processes and alloys |
| 4. Computers | 2. | Radio electronics |
| | 3. | Jet propulsion |
| 5. Lasers | 4. | Automation; remote control |
| 6. Space | 8. | Power equipment; heavy machinery |
| 7. High energy physics | 12. | Natural sciences – basic theories |
| 8. Genetic engineering | | |

$\Sigma$ 108 key projects.

PLATE 4. The transportation sector has been singled out for rapid modernization which includes dredging the Yangtze estuary, deep-water berths for ships up to 100,000 tons and containerization. Coal is still the key element in Chinese energy balance and much is transported by rail. The railway system, covering 50,000 km of main lines and with 8000 steam-engines out of a total of 10,000, will require more electric and diesel locomotives to replace the ageing steam-engines. More and bigger wagons, improved and more double tracks and modern signal systems are also needed within the next few years.

the supply of chemical fertilizer in the earlier plans having mainly been attained and the ambitious Yellow and Yangtze rivers control project indefinitely postponed. Energy comes as a second priority, which is not surprising given the expected needs arising from rapid industrialization and mechanization of agriculture. It is worth noticing that the development of atomic energy for peaceful use was a key element in the 1956 plan; so far China has — for obvious reasons — allocated resources mainly on the defence side. The emphasis on materials and electronics is found in both plans. Lasers, space, genetic engineering, and high-energy physics reflect the importance of new technologies which have come to the forefront since the earlier plan.[4]

However, the economic readjustment to be carried out 1979—1981 has also affected the science and technology sector. The eight priority areas of the 1978—1985 plan were originally planned to have been developed in parallel; the emphasis will now be on the first four. The changes in high energy physics illustrate this. The 40 billion electron volt accelerator originally scheduled for completion in 1983 will be postponed till 1985 and the much larger accelerator has been indefinitely postponed. Similarly, the goal of having 800,000 research personnel in 1985 is likely to have been reduced to less than 600,000.

[4] For a more detailed comparison it is necessary to study the officially released documents which include Fang Yi's speech at the National Science Conference, reprinted in an edited version in *Beijing Review*, No. 14, 1978 "Outline national plan for the development of science and technology, relevant policies and measures", pp. 6—17. The earlier plan is discussed in some details in R. P. Suttmeier, *Research and Revolution. Science Policy and Societal Change in China*, Lexington, Mass., Lexington Books, 1974, 188 pp.

# 4. Policies

AT THE National People's Conference early in 1978, Premier Hua Guofeng in one of his speeches put particular emphasis on developing science.[1] He said that modern science and technology are going through a "great revolution" which will lead to the emergence of new industries and speed up technological development. That advanced science is very much in the minds of the Chinese planners is evident when he says that "modern science and technology are characterized mainly by the use of atomic energy and the development of electronic computers and space science". He also said that China will give full attention to theoretical research in natural sciences including such subjects as mathematics, high-energy physics, and molecular biology, and that there would also be a national research plan for philosophy and the social sciences. The changing views on science and technology are also reflected in the new constitution which has a special article on the subject.

> The State devotes major efforts to developing science, expands scientific research, promotes technical innovation and technical revolution, and adopts advanced techniques wherever possible in all departments of the national economy. In scientific and technological work we must follow the practice of combining professional contingents with the masses, and combining learning from others with our own creative efforts.[2]

The objectives of narrowing the technology gap and increasing labour productivity are themes that transcend the re-awakened attention now given to the development of technology and science. The focus is on the need for new technology in the modern sector where national integrated systems and large-scale manufacturing become logical and unavoidable choices. The policy choices already taken along this road are likely to have two important consequences. First, the highly trained professionals — elite groups or intellectuals — are coming back in their own right, associated with requirements for a certain degree of centralization and hierarchical structures. There is also an urgent need to increase their numbers — that is to enlarge the capacity for turning out professionals within Chinese society by expanding university and postgraduate training, and apparently also a need to have large numbers trained abroad. Second, the technology for the required large-scale, complex systems can, in the short run, only be created to a limited degree within the country.

## Industrial Structures and Technological Needs

To understand better how technology is created and diffused within Chinese industry, we must look more closely at the structure of industrial production that has gradually evolved in China. It is helpful to distinguish between three different modes of production.

[1] Premier Hua stresses importance of developing science (Beijing, Mar. 7, 1978), Xinhua News Agency (Stockholm), 1978, No. 59.
[2] Article 12 of the Constitution of the People's Republic of China, adopted on Mar. 5, 1978, by the Fifth National People's Congress of the People's Republic of China at its first session: BBC FE/ 5759/C/1.

These are differentiated quite sharply here for purposes of underlining the distinct character of each, although, in fact, categories often overlap. The three modes are ranked in descending order of technological sophistication:[3]

(1)  Scientific laboratory industry.
(2)  Urban industry:
    (a) centrally controlled large-scale basic and military industry;
    (b) province and municipality controlled medium-scale industry.
(3)  Rural industry (at county, commune, and brigade levels).

The workshops in the first category — including their scientists and institute sponsors — are predominantly oriented towards achieving self-reliance in high technology, exemplified by computers and electronic components like integral circuits. Their interest in foreign technology has in the past been limited to scientific information, technical literature, scientific exchange, specialized instrumentation, and so forth. Quantitatively, the sector stood for a limited claim on foreign technology. China may now have entered a new development phase when foreign technology and imports will be important even in this sector.

Both the large-scale central plants and the smaller-scale provincial and municipal plants are major users of foreign technology. The potential here for foreign technology extends across the entire spectrum from critical materials that lie beyond China's technical capability to produce high-performance end-products that are urgently needed for the priority tasks of the Chinese economy, sophisticated equipment obtained as single prototypes for copying, and, most importantly, imports of complete plants to boost output in key industrial sectors.

The significance of the rural industrial sector has until quite recently been overlooked by most foreign observers. This is no longer possible with around 20 million people (1978) employed in collectively owned, i.e. commune- and brigade-run enterprises, and at least another 6 million people in state-owned enterprises located in rural areas. Thus a total of at least 26 million Chinese work in rural enterprises, which amounted to at least 40% of China's industrial labour force in 1977 (Table 3).

TABLE 3. *Commune- and brigade-run enterprises in China*[a]

| Year | Number of units (million) | Value (billion yuan) | Employment (million) |
|------|---------------------------|----------------------|----------------------|
| 1973 | 0.5 | 20–25 | 12 |
| 1974 |     | 22.8  |    |
| 1975 |     | 25.1  |    |
| 1976 | 1.09 | 30.1 |    |
| 1977 | 1.39 | 39.1 | 20 |
| 1978 |     |       |    |

[a] J. Sigurdson, *The Changing Pattern of Inter-sectoral Technological Linkages in the Rural Machinery Industry in the People's Republic of China,* country report prepared for International Labour Office, Feb. 1979.

[3] H. Heymann, Jr., *China Approach to Technology Acquisition.* Part III:- *Summary Observations,* 72 p., Santa Monica, *ca.* 1975 (R–1575–ARPA) (a report prepared for Defense Advanced Research Projects Agency), Rand Corporation.

PLATE 5. The manufacture of farm machinery is being stepped up to meet the goals of mechanization of agriculture set for 1980 and 1985. Many of the enterprises are still comprehensive and relatively small units run by communes and counties. A shortcoming of the past has been a lack of machinery to be used in combination with tractors.

These industries manufacture a very wide spectrum of products ranging from chemical fertilizers and plastic goods to diesel engines and machine tools, farm machinery of all kinds, to clothing and footwear. The small enterprises have recently expanded at a much quicker rate than the economy as a whole, which is evident from the following information released from the Xinhua News Agency:[4]

> ... the commune- and brigade-run small enterprises have quickened their pace of expansion in the past two years. In Jiangsu, Hunan, Jiangsi, Guangdong and 11 other provinces and municipalities, total output value of the 800,000 small enterprises shot up 20% in 1976 and another 49.6% in the first 6 months of this year (1977).

Some parts of the country, where rural industries have been less developed in the past, have experienced a very rapid expansion. It is reported from Sichuan — China's most populous province — that the production value in commune- and brigade-run enterprises in 1977 increased by 80% over 1976 and it was expected that the increase in 1978 would be double that of the preceding year.[5]

4 China's commune-run small factories have great vitality, Xinhua News Agency (Stockholm), 1978, No. 3.

5 The Sichuan Conference on Rural Enterprises (Chengdu radio, Provincial Service, Dec. 9, 1978); BBC SWB FE/5993/B11/7.

Against this background we can expect that the rural industrial sector will continue to expand and that the growth rate of the sector — in terms of production value and employment — is likely to exceed that of the modern industrial sector and agriculture where the officially announced targets for annual increases in production value are 10% and 4% respectively.

## Walking on Two Legs

The need for scientific research arises mainly in the modern sector to meet the following demands:[6]

(a) support for defence technology;
(b) support for civilian industrial technology;
(c) cultural pursuit;
(d) long-term strategy to support a fully modernized society and enhance national prestige.

Domestic research capability is of relatively higher importance in the defence sector than in other parts of the economy because advanced defence technology is definitely science-based. Furthermore, China has, since her collaboration with the USSR in the 1950s, chosen not to become dependent on outside suppliers. Recent interest in acquiring foreign weaponry may indicate a shift in this policy.

However, China's science activities play an increasingly important role in agriculture and industry. The traditional view of scientific and technological development emphasizes a linear progression from basic research to innovations, to applied research, and the final application. This view may lead to confusion when we try to understand technological development in China today — particularly within agriculture and small-scale industry. Here, everyday, small, incremental innovations are likely to be much more important for improving production than break-through innovations based on theoretical research carried out over a number of years.

Outside technology sources have not always been available for China's development of her industrial sectors. Worsening relations with the Soviet Union and an embargo by the United States and its allies on what they term "strategic" goods have at times provided great obstacles to importing technology. Furthermore, China has occasionally interpreted her concept of self-reliance to mean that the need for technology resources should be more or less completely met within the country in order not to sacrifice her independence or her domestic political development. China's policy of walking on two legs produced a number of the concrete instruments required for pursuing a policy of self-reliance.

The two-leg policy was originally introduced in order to lessen China's dependence on

[6] Science activities, it should be noted, cover all the following fields:
  1. Research:
     (a) pure research:
     (b) applied research.
  2. Development.
  3. Testing and standardization.
  4. Training.
  5. Information:
     (a) scientific and technical information activities;
     (b) collection of general purpose data;
     (c) popularization of scientific and technical information.

PLATE 6. China has now roughly 500,000 tractors and 1.5 million 10 hp power tillers (1978). Tractors are produced in eight major works. The First Ministry of Machine Building is reorganizing the manufacture of agricultural machinery by forming a limited number of industrial corporations which were originally introduced in 1966. The motives are a desire to utilize economies of scale through specialization, achieve interchangeability of parts, and improve quality – among other things. The picture is from the "October Tractor Plant" in Xinjiang.

the industrialized countries through mobilizing her own domestic resources for rapid industrialization. However, the great hopes of the late 1950s only partially materialized, and the importance of the two-leg policy may today have shifted to reducing the differences between urban and rural areas. But industrially, the policy is still pursued with success in a number of sectors such as cement and chemical fertilizers. Education is also a sector where, for different reasons, a two-leg policy is being implemented, in particular since the overthrow of the Gang of Four.

Since the late 1950s China has more or less systematically implemented a two-leg policy for development. There are five principles: develop industry and agriculture simultaneously; develop heavy industry, light industry, and agriculture but give priority to heavy industry; develop enterprises of all sizes; develop national and local industries; and use modern and indigenous methods of production (Fig. 1). The successful implementation of these policies requires considerable decentralization; but planning, "proper" division of labour, and co-ordination under centralized leadership have always been considered necessary.

In the area of technology, one leg comprises modern industrial technology, much of it imported originally from the USSR, Japan, and Europe. The other leg comprises the technology used in the small scale, generally more labour-intensive units, mostly found in the rural areas. These serve the local areas and may also be part of the country's overall defence system if the big cities should be attacked. Consequently, even if they are shown to be inefficient in comparison with the large modern enterprises, they may still represent

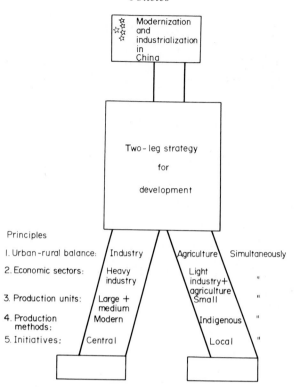

FIG. 1. *Elements in the two-leg strategy.*

fully rational choices to the planners in China. However, China's initial experience in applying the two-leg policy failed when the development of small-scale plants — during the Great Leap Forward — occurred at the expense of modern, efficient industrial production. Furthermore, it was then unrealistically assumed that indigenous small-scale technologies existed in all industrial sectors and that a substantial portion of the rural labour force could be transferred from agricultural to industrial production without any detrimental effects on agriculture.

The concept of a two-leg policy may be confusing when we look at the technologies utilized in China today. In a number of sectors — e.g. petrochemicals, iron and steel, etc. — it is rare to find anything except modern large-scale plants. In many other sectors, however, it is common to find a full spectrum of sizes and technologies from the very small and indigenous to the very large and modern units. Such characteristics have consequences for the R & D structure serving industry. But we must not forget that agriculture and its technological development cover a wide spectrum although the sector may in the future increasingly lean on the large-scale modern leg.

Today agronomy research institutes exist at national, provincial, and prefectural levels, while counties and lower levels usually have mass-based groups responsible for scientific experiments. In combining the professional expertise with the mass-based scientific experiments, it has reportedly been possible to shorten the time needed to produce a new rice strain from 8—9 years to little more than 5 years.[7]

[7] "Mass agro-scientific experiments", *Beijing Review,* 1975, No. 8, p. 23.

PLATE 7. The petrochemical industry is a priority sector which has been developed since the mid-sixties. The picture shows the manufacture of plastic covering to be used in agriculture (a plastics factory in Guangxi Zhuang Autonomous Region).

Hebei Province, for example, has 164 agro-science research institutes with a personnel of about 2000, which is then complemented by the mass-based agro-science research teams engaging on a part-time basis around 1 million people. (Full-time equivalents would be in the region of 40,000.)

The many educated young people, 17 million altogether (1978), who were resettled in rural areas will in all likelihood become an important force in the further development of the local technology system. Such views were clearly stated in a provincial broadcast in Anhui Province in 1977:

> The four-level agro-science network in the countryside, in particular, must absorb the educated youths and bring their role into full play. All professions such as education, science and technology, publishing, and others must show concern for and support the masses of educated youths in their scientific experiments. We must run correspondence courses, amateur education, and all kinds of professional training classes for them and provide them with all kinds of necessary materials for study and equipment for use in experiments.[8]

China's development in the late 1970s is still based on a two-leg strategy with a spectrum of enterprise sizes, technical sophistication, and administrative control. Foreign techno-

---

[8] Hefei, Anhui Provincial Service, Oct. 23, 1977; BBC FE/5658/B11/9.

logy can easily and without much modification be transferred to one end of the industrial spectrum. In other words, the foreign technology is on the whole consistent with the Chinese organization that exists in large centrally controlled, high-technology industries.

Foreign techniques are much less consistent with the organization at the other end of the spectrum. This also applies to Chinese techniques — the difference being that foreign techniques in a certain sense are much more efficient. Consequently, the reasons for the rejection of some foreign technology have to be found elsewhere.

First, the Chinese with their historical experience of the West and other colonial powers want to maintain their initiative and a more or less complete independence of the outside world. Second, efficiency cannot be measured only in classical economic terms. In all likelihood, "less efficient" Chinese techniques include a considerable amount of learning by doing.

It is often claimed that the Chinese leadership vacillated between two different models in organizing research and development. One has been called the mobilization model which aims at the participation of maximum numbers in research. This is based on an assumption that every person is able to make scientific contributions. The other model is characterized by strict organization and professional elites, and is similar to that existing in most industrialized market economies — even granted that it is the Soviet Union which has strongly influenced China. However, taking into consideration China's economic structure and the demands on the R & D system, it appears that China is not in a position to choose a single model for R & D.

Science and technology policy cannot in China — nor in most other big developing countries — be based on a single approach. The needs of the various parts of the economy, or rather the society, will lead to incompatibility if they are lumped together. This relates to scientific knowledge as well as the institutions which generate and distribute knowledge. The pre-occupation with having one model will constantly produce pressure for change to an alternative model. The Chinese leadership has on two occasions attempted to use the same strategy for all sectors of the economy, and this has apparently been a failure. China is today, it appears, formulating a technology and science policy which is made up of various approaches and mechanisms which, however, are constantly influenced by changing domestic and international factors.

## Self-reliance[9]

Self-reliance is an honoured concept and naturally comes to the forefront when discussing industrial and economic planning in developing countries. It stands for self-confidence and reliance primarily on one's own human and natural resources. But it also means the capacity for autonomous goal setting and decision making. One could argue that a country's development will continue to be dependent unless three activities are organically integrated into the national economy. First, a number of essential goods have to be produced within the country. Second, capital accumulation has to be at a level at which foreign development assistance is no longer required. Third, self-sustaining technical progress should be carried out within the country. It appears that the first two conditions are partially fulfilled in China today but the third only to a much lesser extent.

[9] This section is partly based on valuable insights gained from reading *Technological Dependence/ Self-reliance: An Introductory Statement,* by Onelia Cardettini, prepared for the Science and Technology Policy Instruments Project, Apr. 1975.

Self-reliance is conditioned by two major factors. First, self-reliance is primarily a function of the economic and political structure of the country. Second, it is also a function of the size of the country because only if manpower and natural resources are available on a sufficiently large scale can an independent technological and industrial base be established. Naturally, all developing countries cannot apply the Chinese concept of self-reliance. Most countries stress self-reliance to mean only the elimination of special forms of external assistance. However, the Chinese have presented, in rather general form the following interpretation of self-reliance:[10]

> Self-reliance means this: rely mainly on the strength of each country's own people, give full play to their talents, gradually eliminate the forces and influences of imperialism, colonialism and neo-colonialism, and exploit all available resources at home systematically and in a planned way: make every effort to accumulate necessary funds for development through internal sources; take effective measures to train the country's own technical and managerial personnel; in a way suited to local conditions, gradually change the "single-product" economy resulting from a long colonial rule, lift the country from dependence upon, and subordination to, imperialism and establish an independent and relatively comprehensive economic system.

On the questions of national goal and technology transfer, Denis Goulet, in a recently published book,[11] points out that technology is critical to development for the following four reasons. First, technology is a resource and the creator of new resources. Second, it is a powerful instrument of social control even when it reduces underdevelopment. Third, it has a distinct bearing on the quality of decision making in order to achieve social change. Finally, it is a key area where new approaches are needed in order to counter alienation — a characteristic of affluent societies. In sum, technology is in the process of creating new values and by consequence is also a destroyer of old values. Thus a developing country desiring to control its social and cultural development also has to control technology transfer and technological development.

In China self-reliance has been practised at two distinct levels — local and national. In either case it would confuse the understanding of self-reliance if it were reduced to a formula for economic relations alone. Self-reliance practised at these two levels is conditioned by external factors and contains in both cases a partial rejection of the outside goal structure. Thus, at the national level China would in the past reject the goals of the industrialized countries including that of the Soviet Union. Similarly, the rural localities, in practising self-reliance, partly reject the goals of the Chinese cities. Thus China has in principle much more control over external factors in the case of local self-reliance than is the case at the national level.

Self-reliance in rural modernization has in the past come to mean that a locality — team, brigade, commune, or even county — has basically to raise its level of development and standard of living through mobilizing locally available resources and investing the surplus in new projects. Machinery and technology from the outside have mainly been available as catalytic agents — similar to the role played by imported technology in the modern sector. The consequence is that the level of investment and the allocation of manpower are policy issues which are basically left to the locality to decide. However, local development is dependent on a number of critical inputs, and the political need today to speed up rural modernization clearly indicates that larger amounts of inputs

[10] *Beijing Review*, 1972, No. 28 (July 14), p. 16.
[11] D. Goulet, *The Uncertain Promise — Value Conflicts in Technology Transfer*, IDOC/North America and Overseas Development Council, New York and Washington DC, 1977.

might be available, a parallel to the present increasing role of imported technology in the large-scale sector. This no doubt reduces the self-sufficiency of rural areas and may even indicate a re-interpretation of the concept of self-reliance. The same is, of course, true for China's self-reliance on the international scene.

In practising self-reliance at the local level the Chinese leadership is facing a number of problems. First, it is difficult to reduce regional inequalities which are due to different availability of resources, different ability to mobilize creativity among the masses, and different levels of mobilization of the population in general. This may then create a new vertical distinction between self-reliant and non-self-reliant units. Second, self-reliance at the local level may cause exploitation to develop — country by town and poor communes by rich communes — if the basis remains unchanged. This may be the case in places where genuine mass involvement is lacking.

Self-reliance would, of course, lose much of its significance if it were associated with a low rate of growth. Consequently, it must be a gradual process where the rate of growth is accelerated while at the same time developing the capability to sustain it essentially from domestic resources. This is also obvious from an article written by a Chinese economist in 1966 where he summarized self-reliance in the following five points:

> First, to rely on the labour enthusiasm and collective wisdom of the masses of its people in building up the national economy.
>
> Second, fully to utilize all its own national resources (including capital) and base construction squarely on its own manpower, material, and financial resources.
>
> Third, to develop an independent and diversified economy in accordance with its own needs and possibilities, and so gradually to achieve industrialization.
>
> Fourth, to adopt methods that are in accord both with objective laws and with its own specific conditions, and not blindly to copy everything foreign.
>
> Fifth, to develop production, practise thrift, oppose all waste, and gradually to raise the material and cultural standards of its people on the basis of the development of production.[12]

Without making a deeper analysis here it appears obvious that China's self-reliance does not mean total economic independence. There is obviously scope for international transfer of technology, to facilitate the growth of the country's economy even if the main thrust is an adaptation and generation of domestic and local technology.

Cardettini has pointed out that the concept of self-reliance appears to have originated in the 1930s at a time when the areas in China liberated by the communists could not trade freely with the rest of the country.[13] More recently, China's self-reliance has often been viewed abroad as a means of protection arising out of necessity. The emphasis still given to self-reliance — in international relations as well as domestically — is a clear indication of the fact that the Chinese leadership judges that the basic balance of power, which forced it originally to protect its independence, has not changed.

China has since the early 1950s been committed to a global, long-term policy of technological self-reliance. This has at times had very strong political overtones — strongly influenced by the split with the Soviet Union in 1960 which left China for a number of years without a major external supplier of advanced technology. However, China has implemented strict technological self-reliance only as a limited policy in industrial sectors, which from the defence point of view were considered of outstanding importance. It could be argued here that China would have found no willing suppliers of defence techno-

---

[12] "Self-reliance has proved itself", Yung Lung-kwei, *China Reconstructs,* Apr. 1966.
[13] It was first mentioned by Mao Zedong at this time. See Mao Zedong, *Selected Works,* Vol. 1, pp. 141, 170, and following.

logy after the Soviet Union withdrew its assistance, and consequently she had no other choice.

The People's Republic of China has on a number of occasions surprised the world, for instance by the explosion of an atomic bomb in 1964 and the launching of a satellite in 1970. Similarly, her achievements in aeronautics and computers have been quite substantial.[14] It has been shown that China has generally succeeded in reducing the time lapse between the various steps in comparison with the advanced industrial nations, and the examples examined by Macioti have shown that China's scientific and technological progress has been truly noteworthy (Table 4). However, in 1977 Chinese sources have indicated that progress in recent years had been less remarkable.

Very few data are available which show the gap in technological development between China and the major industrialized countries aside from the articles by Macioti, which clearly indicate a dynamic catching up in a few selected sectors. However, in a recent book on China's energy, which will be briefly referred to here, Vaclav Smil has attempted to indicate the level of selected energy technologies.[15]

He says that the mechanization of coal extraction remains low in comparison with the major industrialized coal producers like the United States, Great Britain, and the USSR. More than one-half of all coal is still extracted, loaded, and conveyed non-mechanically, while these operations are almost 100% mechanized in the USA and the USSR. He also points out that oil exploration and production have distinct gaps. This clearly explains the Chinese eagerness to obtain German coal-mining technology and US exploration technology.

None of the principal energy transportation technologies is well advanced in the People's Republic of China. For example, the largest oil tankers built in China are still

TABLE 4. *China closing the technological gap*[a]

| Programme | Year | | | | | |
|---|---|---|---|---|---|---|
| | USA | USSR | Britain | France | Japan | China |
| **Nuclear** | | | | | | |
| First reactor | 1942 | 1946 | 1947 | 1948 | | 1956 |
| First A-bomb | 1945 | 1949 | 1952 | 1960 | | 1964 |
| First H-bomb | 1952 | 1953 | 1957 | 1968 | | 1967 |
| **Space** | | | | | | |
| First satellite | 1958 | 1957 | | 1965 | 1970 | 1970 |
| **Aeronautics** | | | | | | |
| First jetplane | 1942 | 1945 | 1941 | 1946 | | 1958 |
| First Mach 2 jet | 1957 | 1957 | 1958 | 1959 | | 1965 |
| First 8000 kg engine | 1958 | 1957 | 1957 | 1966 | | 1970 |
| **Computers** | | | | | | |
| First prototype computer | 1946 | 1953 | 1949 | | 1957 | 1958 |
| First commercial use of computer | 1951 | 1958 | 1952 | | 1959 | 1966 |
| First transistor | 1952 | 1956 | 1953 | | 1954 | 1960 |
| First integrated circuits | 1958 | 1968 | 1957 | | 1960 | 1969 |

[a]M. Macioti, "Scientists go barefoot", *Successo*, Jan. 1971.

[14] M. Macioti, "Scientists go barefoot", *Successo*, Jan. 1971.
[15] V. Smil, *China's Energy*, Praeger, New York, 1977.

very small compared with the average size of a European tanker currently in service. Extensive construction of pipelines has only just started. Still more important, coal handling is not at a high level of development, although coal constitutes a large proportion of total transportation.

Smil also mentions that the development of high-voltage transmission lines has not been remarkable, and indicates that the upgrading of maximum voltages is lagging even though the necessity for high-voltage direct-current links has already been stressed. Similarly, gas-turbine technology in China is only in its initial stages of development.

Smil considers that power generation is an area where China has been most successful in narrowing the technology gap. This is partly a reflection of the transfer of power technology from the USSR and Czechoslovakia in the 1950s. The Chinese engineers are today able to design and manufacture thermal and hydro-generators up to 300 megawatts. Smil points out that this achievement is quite respectable, but China is "still far behind the current and impending Western and Soviet levels, especially in thermal generator capacities". However, China has been a pioneer in development of direct water cooling for hydraulic and thermal units. Smil mentions that: "They constructed the world's first 12-megawatt steam turbogenerator with inner rotor and stator water cooling in 1958 and introduced 300-megawatt water-cooled sets, both hydro and fossil fuelled, in 1971–4."

Self-reliance is not an objective in itself but a means towards economic independence. Here it may be useful to remember that the Chinese leadership makes a distinction between independence in the political sphere and independence in the economic sphere, as also pointed out by Cardettini in her paper. Economic independence is essential both in the relations between the rural and the urban areas and in China's relations with the external world. It also includes the development of local scientific and engineering capabilities and the adaptation of technology to local requirements.

Self-reliance cannot be equated with self-sufficiency in Chinese economic planning even if we can identify policies that are directed towards self-sufficiency. Generally, the following quotation represents fairly well the policy that China has followed:

> Our self-reliance does not mean closing the doors to the outside world. It is necessary to import some equipment and introduce some techniques from abroad according to the needs in socialist construction with the purpose of increasing our country's ability to rely on herself . . . Learning from foreign countries must be combined with a spirit of independent creation.[16]

Restrictions imposed upon China from the outside, however, have meant that China has accepted, or attempted to follow almost completely, the principle of self-sufficiency in certain sectors. This has in the past applied to defence production. It has also been true for those civilian sectors where the industrialized countries have maintained an embargo on what is termed "strategic goods". Occasionally, China's politicians and planners have attempted to interpret self-reliance to mean self-sufficiency, as during the Cultural Revolution. However, even under such circumstances it has been realized that self-sufficiency cannot be achieved in all production sectors unless China is willing to pay a very high price in developing domestic resources and delaying modernization. This dilemma will be illustrated by referring to the Chinese debate on technology acquisition.

---

[16] "Self-reliance is a question of line" <u>Li Hsin</u>, *Red Flag*, No. 4, Apr. 1, 1975 (translated in *Selections from People's Republic of China Magazines,* CMP–SPRCM–75–13).

**Technology Acquisition**

China started again in 1972 to import complete sets of equipment, this time to be used in producing chemical fibres and fertilizer. She began in the following year to export crude oil to a number of capitalist countries. Deng Xiaoping was apparently one of the advocates of this policy. He was said to have stressed the need for importation of foreign techniques and equipment in order to speed up the technical transformation of industry and raise labour productivity. He is also said to have proposed that the implementation of this major policy should include the signing of long-term contracts with foreign countries. China would then get access to the latest and best equipment which would be paid for by mineral exports. The opponents saw this as "national betrayal" and claimed that:

> ... guided by the principle of independence and self-reliance, it is necessary to import some foreign techniques and equipment on the basis of equality and mutual benefit and in accordance with the needs of China's socialist revolution and construction. But the Chinese people absolutely cannot rest their hopes of realizing the modernization of our national economy on imports. If we do not rely mainly on our own efforts, but (as Deng Xiaoping advocated) rely solely on imports of foreign techniques, foreign designs and technological processes, and imitate foreign equipment, we shall trail behind foreigners and the country's technical development of the whole national economy will gradually fall under the control of foreign monopoly capital.[17]

In the article we learn that Deng Xiaoping considered that the policy had three important advantages as it would enable China to export, to promote technical transformation, and to absorb labour power. On this the opponents had the following to say:

> This means nothing less than opening the gates wide to foreign monopoly capitalists, who would use money and equipment to plunder China's natural resources and suck the blood of the Chinese people whom they regard as low-cost manpower. The Chinese people have had more than enough of such advantages before liberation. If this capitulationist "major policy" of Deng Xiaoping's is followed, China will be reduced step by step to the level of a raw material supplier for imperialism and social imperialism and an outlet for their commodities and investment. This fully reveals the reactionary features of Deng Xiaoping, who works as a comprador of imperialism, representing the interests of foreign big bourgeoisie.[17]

There can be no doubt that an important means by which an industrially backward country can overtake an industrially advanced country is the adoption of advanced technology. The high technology base in China is still quite narrow, a fact which has been pointed out by foreign observers and admitted by the Chinese. For example, in order to speed up the development of the petrochemical fibre industry and thereby contribute to solving the clothing problem of the people, several complete sets of equipment for the production of chemical fibre to be based on petroleum products were imported.[18] In this way China could within a relatively short time transform the petrochemical fibre industry.

In the article referred to earlier, the group from a Designing Institute said that China must draw on the experience and technology of foreign countries. On the character of technology and technology imports they said:

---

[17] *Red Flag* article on Deng Xiaoping's comprador bourgeois economic ideas, Xinhua New Agency (Stockholm), 1976, No. 182 (Xinhua and Beijing, July 30, 1976).

[18] Denouncing the absurd theory of the Gang of Four on the question of introduction, by the criticism group of the Second Designing Institute under the Ministry of Light Industry, *People's Daily*, July 3, 1977; CMP–SPRCP–77–28.

The advanced technology of foreign countries is created and accumulated by the people through long production practice, and constitutes a part of the wealth in the treasure house of human knowledge. The bourgeoisie can make use of them to develop capitalism, and our dictatorship of the proletariat, with its superior socialist system, should make full use of all advanced scientific and technical gains to build socialism.

... Technology and equipment are introduced without any political strings attached, without bartering away our sovereignty, without carving up our profits, without clauses of a slavish nature, and without the loss of national dignity. How can there be any talk about "fawning" on the foreigner and national betrayal? [18]

From such facts we can see that China experienced a very acute struggle over the question of how and to what extent China should make use of foreign technology. That the issue has been more or less resolved can be seen from the following statement in which the substance of self-reliance is discussed by a leading member of China's planning commission:

We don't practise autarky by locking our doors against the world. We will make positive efforts to have economic and technical exchanges with foreign countries and expand our foreign trade. Following development in industrial and agricultural production, we shall sell more and more petroleum, coal, and other products in order to buy advanced foreign equipment. China does not accept loans from foreign governments, nor do we collaborate with foreign countries in the exploitation of our resources. In some areas of economic exchange, China will continue to adopt the usual international practice of deferred payments. Advanced techniques and experience are the common wealth of the working people throughout the world, and China will make efforts to learn and master them. [19]

Since the statement was made, China has gradually come to modify its views both on foreign loans and joint projects.

The usefulness of importing plant and machinery becomes obvious when related to the ambitious modernization plans which were spelled out in some detail at the Fifth National People's Congress held in February 1978. At the time, Premier Hua Guofeng mentioned that China in the next 8 years plans to build 120 large-scale projects which would include, among other things, 10 iron and steel complexes, 10 oil- and gas-fields, and 6 new trunk railways. [20] While the emphasis is on large-scale projects, the Premier stressed that " ... every attention must also be given to the development of medium-scale and small enterprises".

The Premier also said that China would by the year 1985 produce 60 million tons of steel — against approximately 25–30 million in 1978 — and that the annual increase in industrial output value would exceed 10% in the coming 8 years. In describing the industrialization policy, he also pointed out that China will develop a regional economic system in six major regions, and that the modernization drive will provide China with "14 fairly strong and fairly rationally located industrial bases". [21] (The composition of the regions and their varying resource bases is given in Appendix III.) In addition to the development of the basic industries, the Premier also stressed that China would have "much more developed petro-chemical, electronics and other new industries" and that the country "would build transport and communications and postal and telecommunications networks big enough to meet growing industrial and agricultural needs". [22] Foreign

[19] Leading member of planning commission on China's modernization, Xinhua, Dec. 31, 1977, Xinhua News Agency (Stockholm) Jan. 3, 1978.

[20] Chinese Premier announces building of 120 large-scale industrial projects (Beijing, Mar. 6, 1976), Xinhua New Agency (Stockholm), 1978, No. 58.

[21] BBC FE/5757/i.

[22] Premier Hua Guofeng puts forth 1985 economic norms (Beijing, Mar. 6, 1978), Xinhua News Agency (Stockholm), 1978, No. 58.

technology has consequently become more important, and Shannon Brown, for example, argues that:

> Because importation is the most efficient method of acquisition, nearly all new technology introduced in China in the near future will originate abroad and will be brought into the country by ... the importation of printed matter, machinery and equipment, and complete plants, and the transfer of knowledge through the movement of people. Very little technology is likely to be developed *de novo,* although considerable adaptation of imported technology to Chinese conditions will be necessary.[23]

Naturally, this means a different interpretation of self-reliance. Brown argues that self-reliance will increasingly mean the "mobilization of Chinese efforts to select, adapt, disseminate, and use within China the highly productive new technology already developed elsewhere. China's efficiency and effectiveness in performing this task will do much to determine her rate of economic development.".

If we accept recent Chinese statements on the country's level of technological development, and there seems to be no reason for discounting them, a picture emerges which shows increasing technology gaps in a number of industrial sectors. The machine-building minister responsible for electronics has said that the gap in many areas of electronics has been widening.[24] The Minister of the Coal Industry expresses a similar concern and says:

> that in the early 1960s the gap in mechanization between China's leading coal-mines and the world's major coal-producing countries was not too large. However, it was widened later on. China is behind advanced-world levels in labour productivity and other economic and technical indicators.[25]

The same concern about technology gaps is also evident in an interview with Qian Sanqiang, deputy general secretary of the Academy of Science (Academia Sinica). He stated to Tanjug News Agency "that compared to the level of science in the world today, China is between 10 and 20 years behind, varying from field to field. This gap was smaller in 1965, but it increased during the Cultural Revolution . . . "[26]

The close link between importation of technology and Chinese exports can be seen in the 8-year "private" trade agreement that China signed with Japan.[27] Under this China will import $US 10 billion of plant and machinery during a 3-year period beginning April 1, 1978. Included in this agreement may be an integrated coastal steelworks with an annual capacity of 6 million tons to be delivered by Nippon Steel and constructed in Shanghai. China is to pay for her imports from Japan mainly by exporting crude oil and coal. Apparently there is a strong mutual interest in promoting trade whereby technology will flow from Japan in exchange for raw materials, in particular minerals. This is obvious in the newsletter from the Japan External Trade Organization (JETRO) in early 1978, which says that:

> the import items which China wants were listed in definite terms in the contents of the Agreement, e.g. plant related items like coal mining and coal dressing equipment, transport equipment,

[23] S. R. Brown, *Foreign Technology and Economic Growth, Problems of Communism,* Vol. 26, No. 4 (July–Aug. 1977).

[24] Minister interviewed on electronics industry prospects, NCNA, Nov. 16, 1977; BBC FE/5683/B11/10.

[25] Chinese Minister of Coal Industry on mechanization, Xinhua News Agency (Stockholm) 1978, No. 16 (Beijing, Jan. 18, 1978).

[26] Tanjug in English, Jan. 26, 1978 (BBC FE/5730/B11/12:- the quotation marks refer to the text printed in the *BBC summary of World Broadcasts*).

[27] "Machines for oil", *The Economist,* Feb. 18, 1978.

port and harbor equipment, petroleum mining equipment, metallurgical equipment, equipment for the petrochemical industry, equipment for electricity generation, etc., technological materials such as patents and know-how; and building materials with the emphasis on iron and steel. Naturally, these items were very attractive to Japanese industry, currently distressed over the stagnation of business. Moreover, the hope was probably that in the long term this Agreement may lead to the construction of a progressive framework for strengthened economic ties between Japan and China. On the other hand, there were also strong requirements on the Chinese side to rehabilitate the economy and pull it out of the chaos and the destruction caused by the Gang of Four and to achieve as quickly as possible its "four modernizations".[28]

The trade agreement indicates that Japan will remain China's biggest supplier but China will also sign big contracts elsewhere — possibly under the umbrella provided by the 5-year trade agreement with the European Economic Community signed in early 1978.[29]

Given the present maldistribution of the global capacity for research and development, the Chinese policy of stressing the importation of technology of various kinds is apparently a logical choice. The Chinese R & D budget is reported to be 5.9 billion yuan in 1979, which is twice the Swedish allocation for the same purposes — in a country with one-hundredth of the Chinese population. Seen in a global perspective, the Chinese allocation of R & D resources is of the order of little more than a couple of per cent of the world's total spending.

When developing technology and choosing between different alternatives, a developing country also has to give special consideration to the following three aspects. First, technology has become increasingly complex — a complexity which is usually found not only in the production process but also in the product itself and in the distribution system. Second, technology is constantly changing in all these three dimensions. A major reason is that most technology originates in industrialized countries where a demand for continued growth and higher real wages leads to a 5–6% increase in efficiency/productivity annually. Over a period of 5 years this amounts to 30–35% increase in productivity/efficiency. Third, the educational needs are very considerable, which is to a large extent a consequence of the first two considerations.

Two important consequences follow from an assumed policy of increasingly using foreign technology. First, the selection of foreign technology requires increased information about products and processes, and such data must be screened, systematized, and made available to large numbers who can evaluate and criticize independently what is potentially available. The need to follow closely developments abroad is also evident in scientific research. Zhou Peiyuan said that the various disciplines are being knit ever more tightly together with leading fields and new branches of science developing rapidly. With this in view and the achievements abroad, he underlined that vigorous academic exchange must be seen as an integral part of scientific research. So he stressed that "researchers in every project must keep pace with latest developments, both domestic and international".[30] Second, the importation of foreign technology requires foreign exchange, which in the absence of long-term credits or development assistance calls for a commensurate development of the export sector. At present Chinese exports mainly consist of agricultural

---

[28] "The Japan–China long-term agreement", *JETRO China Newsletter,* No. 16 (Jan. 1978).
[29] *Recommendation for a Council Regulation – Concluding the Trade Agreement between the European Economic Community and the People's Republic of China,* Commission of the European Communities, COM(78) 68 final, Brussels, 22 Feb. 1978.
[30] Zhou Peiyuan calls for vigorous academic exchange (*People's Daily,* Feb. 6, 1978), Xinhua News Agency (Stockholm), 1978, No. 38.

produce and minerals. The industrial goods exported are mainly competitive because of price rather than performance. Boosting the export sector would, aside from the policy decision to allocate investment resources, also require new, partly imported technology in order to reduce costs and improve product quality. Otherwise, China would use low-cost labour with low productivity in order to pay for the imported technology used elsewhere in the economy — a situation which might be accepted for a certain period of time. Thus a dilemma is facing the Chinese planners. In order to pay for imported technology it will be necessary to import still more technology.

Many developing countries accept joint ventures in order to get access to efficient foreign technologies. Other countries who want to have a better control of their development reject this but often accept various other schemes. Recent views on the subject in the USSR can be seen from the following quotation:

> Product-payback schemes have also gained wide currency. Under these schemes the socialist countries attract foreign credits for the construction of enterprises and eventually supply their creditors with a share of the products these turn out. Such credits are usually channelled into areas where the STR (scientific and technical revolution) is in full swing.[31]

Until recently there was no indication that the Chinese leadership would accept such methods even in high technology areas, where a collaboration of this sort might be beneficial in a narrow technical sense.

However, new policies on how to finance large-scale imports of foreign technology were apparently discussed during the summer 1978. Aside from long-term credits, China obviously considered product payback schemes, and one of the vice-premiers, Yu Qiuli, had the following to say at the national finance and trade conference:

> China intended to expand foreign trade primarily by its exports . . . more industrial products and minerals and durable consumer goods would be exported, and some surplus equipment and labour used to process raw material, copy prototypes, and extend assembly lines. *Some enterprises that needed particular equipment would be allowed to import it, paying with the products produced. Special factories and areas would be set up according to plan to deal mainly in export goods.*[32]

A product payback scheme is dependent on the demand for the products in the world market, and the present economic situation would indicate that this approach is more viable for certain raw materials than for engineering products in general.

A main incentive for avoiding foreign technology is the shortage of foreign currency. But technology is no mere commodity; it is an overall system. To harmonize acquired technology with the objectives of national development has been regarded by many countries as a major priority. However, many years are likely to elapse before the developing countries can build up an R & D capability with infrastructures of a size, diversity, and strength comparable with what exists in already fully industrialized countries.

The importance of using foreign technology lies in its ability to achieve operational efficiency in a short time. However, this usually requires the simultaneous development of infrastructure like transportation, training, and maintenance facilities, among other things. The separation of the desired components from undesired ones in transferring technology from the advanced capitalist countries may not always be very easy. Everett E. Hagen presents the problem in the following words:

[31] *Socialism and the Scientific and Technical Revolution* (25th Congress of the Communist Party of the Soviet Union), Progress Publishers, Moscow, 1977.

[32] Vice-premier Yu Qiuli on China's foreign trade policy (Beijing, July 2), Xinhua News Agency (Stockholm), 1978, No. 159.

...every Western industry depends for its efficiency on other industries. It assumes the ready availability of materials, components, and tools. It depends also on auxiliary enterprises which can provide technical, financial, and managerial services on demand; on a complex network of communication and transportation facilities; and on an intricate system of business practices. A Western economy is a technical (and cultural) complex, not a set of isolated pieces of technology. In an undeveloped society the auxiliary industries are missing and the framework of business practices is different. One piece cannot be detached from the complex and used efficiently elsewhere without skilful adaption.[33]

We also have to remember that technology is constantly changing because those possessing it are gaining a decisive advantage over others when it comes to economic competition. This is particularly so in certain areas characterized by rapid development of "unstable technologies" of which electronics is a typical example. Here technology is constantly challenged with subsequent changes in products, manufacturing process, and production costs. Such changes obviously have consequences not only for technology as a resource for creating wealth; they also have a direct bearing on the forms of social control and the modes of decision making.

Denis Goulet points out that technology affects development on the following major counts:

It is a major resource for creating new wealth; it is an instrument allowing its owners to exercise social control in various forms; it decisively affects modes of decision making; and it relates directly to patterns of alienation characteristic of affluent societies.[34]

In discussing and understanding these aspects we constantly have to keep in mind that technology is both a system of its own and a component of larger social systems. Consequently, value choices, development strategy, and technology policy are all interconnected, and we must look into the inner dynamics of technology as well as its links with the broader processes in society.

## Open-door Scientific Research

The coupling, or integration, of the research laboratories with the outside world serves two important but very different functions. First, it provides, or rather diffuses, scientific and technical knowledge to lower levels and far-away units. Second, it integrates scientific research with production in order to facilitate application and speed up the development process. We also have to realize that political functions in China are involved in both cases. A successful integration in the first instance enables the narrowing down of differentials between regions and between individuals of various categories aside from its contribution to productivity. In the second case the successful integration enables a quicker expansion of the material base — in order to build socialism — than would otherwise have been the case. In trying to fulfil the two functions it is necessary to draw on the same resources such as manpower and institutional equipment, and it is natural to find conflicts over the priorities.

The number of people participating in the China "research system" through attempts

[33] E. E. Hagen, *On the Theory of Social Change: How Economic Growth Begins*, Dorsey Press, Cambridge, 1962, p. 31. My attention was drawn to this book when reading an article, "Transfer of military technology to Third World countries", by S. Landgren-Bäckström, in *Bulletin of Peace Proposals*, (PRIO) Oslo, 1977, No. 2.
[34] D. Goulet, *The Uncertain Promise — Value Conflicts in Technology Transfer*, IDOC/North America and Overseas Development Council, New York and Washington DC, 1977, p. 7.

to carry out research and make innovations, as well as in technical education and the popularization of technical and scientific knowledge, is most likely much greater in relative terms than in any already industrialized country. This contrasts greatly with the situation in the countryside in most developing countries. Particularly, the scientific invasion of the countryside is bound to stimulate the minds of rural people in new directions, undermine their superstitions, and open up the path to their intellectual emancipation and full participation in the modernization of China.

The idea underlying the open-door policy of the research institutes was a basically good one. It meant that scientific research could no longer be carried out without any concern for the realities outside the laboratories. The services and training provided appear to have been useful in diffusing scientific and technical knowledge. The approach was also essential in establishing local centres for research or spreading and applying new knowledge, the importance of which becomes evident from the information provided in the chapter on mass science. However, the programme tended to drain so many resources from the research institutes that many places had little time and opportunity to do research in other equally important areas. Occasionally, open-door research also meant the transfer of equipment. As some of the shortcomings became evident and changes were suggested to remedy the situation, the "radicals" apparently counteracted by stressing the political unreliability of intellectuals in general, and scientific researchers in particular. Consequently, they should be controlled and remoulded, and open-door research was one of the instruments for this. Other reports were published which indicated that the method had no disadvantage in the use of resources or for reaching scientific goals.

There were several ways of conducting open-door scientific research.[35] The Chinese Academy of Sciences, for example, practised the following three forms. First, it encouraged the researchers to leave the laboratories and involve the people in factories and communes concerned with research problems within the units. Second, it opened up the research institutes and invited people to come there to do scientific research. This also included the encouragement of institute workers to participate in research and management. Third, it also provided open-door service by accepting visitors asking for assistance on research matters, a form which often involved the training of technicians. At its peak the open-door scientific research programmes of the Academy involved several thousand researchers who regularly went to the factories and to the countryside.

The criticism of the open-door research with its emphasis on "the three-in-one combination of leading cadres, experts and masses" does not mean that the programme should be scrapped. However, it is now being stressed that an important condition for doing a good job in this area is that the professional contingent should be the backbone. A further measure should be "to train more working class technicians, engineers and scientists, constantly expand and improve this contingent and make them the backbone of China's scientific endeavours".[36] Administrative changes introduced in early 1979 now allow the institutes much greater freedom to make their own decisions and they are encouraged to engage in contract research for enterprises.

In any country there is always a gap between policy and implementation. An example of how the earlier policies, advocated by the Gang of Four, were not always followed,

---

[35] *Open-door Practice — Correct Way for China to develop Scientific Research*, NCNA, Apr. 4, 1976, Survey of People's Republic of China Press; CMP–SPRCP–76–15.
[36] *Red Flag* article discusses modernization of science and technology in China, Xinhua News Agency (Stockholm) 070505, Beijing, July 5. 1977.

comes from a national quantum chemistry conference, which was the first of its kind organized by the Chinese Academy of Sciences. In a report from this conference we learn that "even during the days when the Gang of Four ran amok, Prof. Tang Ao-ching and the quantum chemistry research group of Jilin University persisted in carrying on research work and advanced a new theory on the principle of symmetry and constancy of molecular orbit(s)".[37]

It may be questioned whether open-door research and the integration of researchers with workers and peasants really had the impact on the scientific research establishments now implied in the criticism against the Gang of Four. In a study of the journal, *Scientific Experiments,* the publication of which was announced in April 1971, S. P. Rawal says that the picture emerging for 1974 is that the coverage given to articles written by workers—peasants—soldiers or to innovations made by such groups is limited.[38] This is still more true for *Scientia Sinica* — the journal of the Chinese Academy of Sciences — where many articles during 1974 carried a plea for renewed interest in research in the basic sciences.

There are two main interpretations of what Rawal and others have observed with regard to the discrepancy between policy and implication in science and technology. One possibility is that the "radicals" were right in stating that the policies — advocated by Mao Zedong — failed because they were thwarted by the intellectuals and other power holders who wanted to reinstitute the system that had existed before. Another possibility and the one underlined by the present leadership is that the new ideas and policies could not be implemented — at least at the present stage of economic development — and consequently the Gang of Four advocating such mistaken beliefs had to be removed.

## The Intellectuals

Looking back provides an improved perspective on the role of intellectuals in China's policies for science and technology. Already in 1949, when the People's Republic of China was established, scientific research and development immediately received careful attention. It was seen as one of the important instruments for achieving industrialization and modernization. The academies, primarily the Academy of Sciences and the industrial research institutes, were receiving large allocations for their expansion. The whole system for science and technology, including universities and colleges, was influenced and inspired by the experience of the Soviet Union which also provided much of the expertise in the early years.

China in the 1950s was still bent on rapid industrialization along Soviet lines. With the Great Leap Forward in 1958 came the inspired belief that modernization could be greatly speeded up. Simultaneously the scientific and technological contributions from laymen were considered equal to those of the experts, in almost all areas. The economic difficulties in the early 1960s, which were only partly due to the policies of the Great Leap Forward, led to a reduced rate of development. At the same time the ideological commitment to build a socialist state, in order to attain communism, was subordinated to eco-

[37] NCNA, Jan. 8, 1978; BBC FE/5712/B11/20.

[38] S. P. Rawal, *Reflection of the General Line in Science and Technology: A Study of* Scientia Sinica *(1974) and* Scientific Experiments *(1974), China Report,* Vol. 12, No. 5 (Sept.–Oct. 1977), pp. 3–15, New Delhi.

nomic consolidation. With the Cultural Revolution in 1966—9, ideology was again brought to the forefront.

The Cultural Revolution aimed its spearhead against the bureaucracy and its inability to handle the country's development problems. The lack of equality and justice among different sectors and groups of people was attacked and this also affected the science and technology establishment. The attacks very soon encompassed all groups that were considered experts or elites. This development was apparently not envisaged as is evident from the comments on scientists and technicians in the circular which was to guide the Cultural Revolution.

> As regards scientists, technicians, and ordinary members of working staff, as long as they are patriotic, work energetically, are not against the Party and socialism, and maintain no illicit relations with any foreign country, we should in the present movement continue to apply the policy of "unity, criticism, unity". Special care should be taken of those scientists and scientific and technical personnel who have made contributions. Efforts should be made to help them gradually transform their world outlook and their style of work.[39]

Demands for being both Expert and Red were high and few at the time were considered to have fulfilled the requirements. The contributions of laymen and mass-based science came to the forefront very much as in the Great Leap Forward. This has to be seen in connection with the increasing attention to agriculture and rural development. This also included improved programmes for public health, education, electrification, and industrialization of the countryside. It is now being realized that the criticism and harsh treatment of the intellectuals were not justified.

From the beginning of 1977 the views of the intellectuals have completely changed. Industry and agriculture are progressing. What should we do with culture and education? What should we do with the scientific research? What should we do with the intellectuals in this mass movement?[40] These questions were asked at the end of a long article which was published in summer 1977. *Red Flag*, the theoretical journal of the Party, discussed the earlier policy towards the intellectuals. In this context, as in many others, the Gang of Four was accused for having carried out wrong policies with the consequence that the intellectuals were pushed aside, and important assets for China's modernization wasted. However, the Gang of Four criticized the intellectuals for considering knowledge as private property. This is now said to have meant a criticism of the knowledge itself as well as the carriers of that knowledge. This in turn might have created a situation where the Chinese people would not have been willing to acquire new knowledge — with disastrous consequences for China's future development. Here is a central issue. The Gang of Four maintained, according to the reports reaching us, that the individuals who acquired new knowledge might become enemies of socialism and, consequently, had to be treated accordingly. There can be no doubt that this group is now being treated differently although mass science and the significance of the non-expert is still being stressed. But, technical and scientific personnel will again be given titles and transferred to places where they can use their ability as well as having more time for their work. But technicians, engineers, and researchers form a comparatively small group — at most a few million.

A number of policies and measures regarding intellectuals are now being implemented

---

[39] Point 12 on "Policy towards scientists, technicians and ordinary members of working staffs". Decision of the Central Committee of the Chinese Communist Party Concerning the Great Proletarian Cultural Revolution (adopted on Aug. 8, 1966), *Beijing Review*, 1966, No. 33.
[40] *Red Flag*, July 1977.

which can be summarized in the following six points. First, the intellectuals are accepted as an important asset and force in the modernization of the country. Second, the principle of letting a hundred schools of thought contend is fundamental in the development of science and technology. Third, technical posts entail specific responsibilities. Fourth, all research institutes practise the system of directors undertaking responsibility under the leadership of party committees. Fifth, an adequate number of work hours for professional work are ensured. Sixth, titles for technical personnel are restored, being part of the incentive system. Already towards the end of 1977 the Chinese news agency reported that with the approval of the State Council the Chinese Academy of Sciences had appointed more than twenty directors and deputy directors of its research institutes.

PLATE 8. Laser technology has been singled out as one of eight priority research areas besides computers, space technology, high-energy physics, material sciences, genetic engineering, energy, and agriculture. China has in the past committed considerable resources and established a good foundation in laser technology. The picture shows scientists treating maize seeds with laser at the Agricultural Science Institute in Xiyang County where the model brigade Dazhai is located (Shanxi Province).

Here it was mentioned that the system of directors of research institutions having responsibility under the political leadership had played a positive role in promoting the sciences — but had been sabotaged in the past. The re-institution of the system is now expected to organize effectively, and direct, scientific research and also stimulate the research workers.

The question of management was clarified by Premier Hua Guofeng at the National People's Congress in stating that the revolutionary committees, after being re-elected, will be maintained at the higher administrative levels — down to communes. However, at the lower levels a system of division of responsibilities will be used. Under this system there will be "factory directors, production brigade leaders, school principals, college presidents, and managers taking charge under the leadership of the party committees".[41]

On the question of incentives and evaluation of research workers, Kang Pai, head of the leading group of the Institute of Physics, had the following to say:

> We are establishing a system for the evaluation of scientific and technical workers, restoring titles for technical personnel, enforcing the system of keeping evaluation scores, and using various other systems including that of personal responsibility and participation by the masses in managerial work.[42]

As an element in the preparations for the coming National Science Conference we learn from the Xinhua News Agency that workers are busy making medals and certificates of citations to be awarded to outstanding groups and individuals at the conference. The office, set up to prepare the conference, disclosed that several thousand scientists, technicians, and other people have been recommended for that honour.[43]

The recent debate on education and on the role of the intellectuals, including those working in the technical and scientific fields, has centred on two issues. One is the question of whether the bourgeoisie or the proletariat exercised control over education in the years preceding the Cultural Revolution. The other question is whether or not the great majority of the teachers and students who graduated in those years were bourgeois in world outlook. Beijing University has the following to say on the question, and similar figures and arguments have been repeated by other universities and colleges:

> In the 17 years after liberation, more than 19,000 students graduated from Beijing University — an average of more than 1000 a year. A great change took place in the class composition of the students. In the early days after liberation the number of university students who were workers and peasants or who came from worker and peasant families was very small. In 1950 Beijing University established a worker–peasant middle school to train worker–peasant cadres speedily and to assist the children of workers and peasants to attend universities. This was the first worker–peasant crash-training middle school established by an institution of higher learning in the country. By 1956, 67% of the total student enrolment came from among the workers and peasants or from worker–peasant families.[44]

The Gang of Four also appear to have claimed that those who received education were likely to turn into intellectual aristocrats and consequently saw mass science as a countervailing force which should be supported by all means. Their views on this matter had, of course, an immediate bearing on the treatment of people of special talent.

[41] BBC FE/5757/i.

[42] Beijing Home Service, Sept. 29, 1977; BBC FE/5634/B11/6.

[43] Beijing scientists and artists look to 1978 with confidence, Xinhua (Beijing) Jan. 2, 1978; Xinhua News Agency, (Stockholm), Jan. 3, 1978.

[44] Beijing University leaders criticize Gang of Four, Xinhua New Agency (Stockholm), 1977, No. 300.

Yang Cheng-chung, a research worker at the Academy of Sciences, writes that "as a result, young scientific workers could not receive excellent training, and the average age of the contingent was higher than normal". He also says that "the Gang of Four tried in every way to attack the young scientific and technical personnel who were aces in their vocational work".[45] In the same article Yang Cheng-chung clearly indicates that talented intellectuals have a special role to play:

> As in the practice of class struggle and the struggle for production, advanced elements will always emerge in the practice of scientific research. This is an objective law of the development of things. The broad masses can make only with the leadership of a few aces further progress and raise their proficiency. On the basis of such leadership, new aces will be produced.

The question of whether to rely on a limited number of experts or to mobilize a large number of working people was singled out as a key issue in *China Reconstructs* (June 1976). The article says, with apparent reference to Deng Xiaoping, that he alleged:

> ... that China's science and technology can develop only by depending on "several hundred first or second-rate scientists", and that "the workers, peasants and soldiers have too low an education level" to qualify them for scientific research. By using the Marxist–Leninist point of view and the real situation, the broad masses have sharply denounced this fallacy. A host of facts show that it is the working people, not a few geniuses, who have created history and science, and that the workers, peasants, and soldiers are the main force in the struggle for production and scientific experiment.

The intellectuals in China are now given more opportunities to make their contribution to China's modernization. From the available evidence this does not indicate that the role of the masses in scientific research will be belittled, nor will it mean that scientific and technical personnel will refuse to solve urgent problems in production. But the confusion over organization and resources policy in science and technology has disappeared.

In their positive evaluation of the intellectuals, the new leadership refers to comments that Mao Zedong made back in 1957. Mao then indicated that 3% of the intellectuals were hostile — that is, towards the Party. There were also hesitant groups, but more than 90% of the intellectuals supported the socialist system — but in varying degrees, Mao said.

At the same time Mao also discussed the dictatorship of the proletariat. He mentioned that in China at the time, the proletariat, which of course excludes the peasants, consisted of 10 million while the intellectuals then numbered 5 million. We now learn that China had 20 million intellectuals in 1971 and that this number is too small considering the modernization programmes. Expansion of education in universities and colleges is required in order to support industrial and technological development — particularly when China starts to rapidly expand new industrial sectors.

In addition there is reason to assume that China will experience a considerable expansion in employment in the administrative units of organizations as she proceeds on the modernization path. In particular we can expect an increase in the number of people responsible for the exchange and processing of information. The neglect of the sector handling and processing information and knowledge (the quaternary sector) — the scientific and technical journals, for example, ceased publication during the Cultural Revolution — is likely to have become more and more serious in recent years and may be one of the major reasons for revising the national policy towards education and the intellectuals.

[45] Yang Cheng-chung, "Our cause requires the aces", *People's Daily*, Aug. 11, 1977; BBC FE/5592/B11/3.

Related to this are the regional and spatial consequences of many decisions. Studies of organizational decisions often reveal that "decisions other than pure locational decisions can have spatial consequences".[46] This seems to have been one of the problems of open-door research policy and the close integration of production with scientific research. The consequence was that the widely dispersed localities and the large number of production units demanded that the limited equipment and number of personnel in the central units should be put at the disposal of those who demanded such resources.

## Information and Publication

The publication of scientific journals and various periodicals covering technical disciplines has been closely related to the policy on intellectuals. In 1966 the number of periodicals officially distributed by the Chinese 'post office amounted to 648. The different categories are listed in Table 5, which shows that more than half of the periodicals were providing information on various aspects of science and technology – a considerable number of them providing abstracts and translations of articles appearing in foreign journals.[47]

The Cultural Revolution gradually led to the discontinuation of the official publication of practically all journals and magazines. At the end of 1967 there may not have been more than 10 journals with a scientific and technical content available. Research and technical development did not stagnate though, even if it was occasionally seriously disrupted. So it is likely that at least part of the flow of scientific and technical information was organized in different ways.

TABLE 5. *The postal administration's categories of newspapers and periodicals 1966*

| | | |
|---|---|---|
| Daily papers | | 12 |
| Politics, current events, and social sciences | | 33 |
| Finance and economics | | 10 |
| Periodicals for youth, children, and women | | 11 |
| Culture and education | | 28 |
| Literature and art | | 75 |
| Sick care, pharmacology, and hygiene | | 63 |
| Science and technology: | | 380 |
| Digest on science and technology | 99 | |
| Bulletins | 8 | |
| Documentation | 83 | |
| Review index | 26 | |
| Express bulletins | 20 | |
| Translations | 77 | |
| Transactions | 67 | |
| Periodicals in foreign languages published in China | | 36 |
| Total | 648 | |

[46] Such problems are discussed at some length in "The geography of economic activities: some critical viewpoints on theory and application", by Gunnar Törnqvist, *Economic Geography*, Vol. 53, No. 2(Apr. 1977).

[47] Swedish Academy, *Science, Technology in Communist China*, report No. 154 by J. Sigurdson, published by the Swedish Academy of Engineering Sciences: Stockholm, Naturvetenskap och teknik i Kina/Natural Science and Technology in China), Swedish, 1968, pp. 1–82, IPRS/L 2972 (Feb. 28, 1969).

Further reflection shows that the supposition is not entirely unreasonable. There are two phenomena that support such an opinion. To begin with scholarship has traditionally been very highly regarded in China. Academic examinations and academic standing have therefore been coveted status symbols. In learned circles practical and applied knowledge has not always been very highly valued. Many of the Chinese scientific and technical journals were criticized for holding on to obsolete academic notions that have made it difficult for laymen researchers to get their contributions published. Extensive editorial work also led to a considerable delay in the publication of research results.

Some of the periodicals that have failed to appear may, of course, for reasons of national security, have been transferred to internal distribution. However, the information media may also have been reorganized to fit the new economic planning. The integration of education, research and production, and the considerable influence of "non-experts" require new forms of scientific—technical information and debate. The new research policy strongly emphasized the fact that research efforts are for the people, i.e. for the national collective. The former publication of journals in China also had to a certain extent the character of an accreditation system, i.e. part of a career system for elite groups rather than serving national needs. This was one of the points of criticism during the Cultural Revolution. Another contributory cause may have been increased demand for speed. Many areas of Chinese research have a strong local concentration. Group discussions and simple mimeographed summaries could then advantageously replace the former journals.

Such a system would probably function well in China where the research establishment is centrally controlled and where the research within each sector is locally concentrated in one or a few places. The development work that builds further on the research results, i.e. construction of prototypes and initiation of trial production, is often placed with enterprises near the institution that was responsible for the original results.

Visitors to China over the past few years have noticed the resumption of scientific and technical journals which are for trial purposes or were only available for internal circulation. Furthermore, technical and scientific information can be distributed in many different ways, and is not necessarily contained in glossy magazines.

Whatever the characteristics of the internal system for distributing and discussing scientific and technological information may have been, there is no doubt that it also contained serious drawbacks. Among the more serious may have been the lack of internal criticism and the lack of access, in many quarters, to foreign data and information on progress abroad.

However, official publications have gradually come back. One of the first to appear was *Scientia Sinica,* the prestigious journal of the Chinese Academy of Science, which resumed publication in 1973. The *Chinese Journal of Medical Sciences* appeared at about the same time. A popular magazine — *Scientific Experiment* — began publication in 1971. The publication of *Archaeology and Archeological Studies* resumed in 1973 after having been reportedly approved by Zhou Enlai. A few years later there were about 25 scientific and technical journals available, and the number early in 1978 has increased considerably. The situation is now in all likelihood the same or better than in 1966. The information folder of the Institute for Scientific and Technical Information of China (Beijing) reports for 1978 that the total number of Chinese periodicals is 2,500 of which 70% deal with science and technology.

Also of significance is the appearance of a journal with the name *Schools of Natural Sciences in Contention,* which was announced in 1977. The publication of such a journal is, of course, related to the re-emergence of the policy "Let a hundred flowers blossom, let a hundred schools of thought contend". On this Mao had the following important comments to make:

> Letting a hundred flowers blossom and a hundred schools of thought contend is the policy for promoting the progress of the arts and the sciences and a flourishing socialist culture in our land. Different forms and styles in art should develop freely and different schools in science should contend freely. We think that it is harmful to the growth of art and science if administrative measures are used to impose one particular style of art or school of thought and to ban another . . . Often correct and good things have first been regarded not as fragrant flowers but as poisonous weeds. Copernicus' theory of the solar system and Darwin's theory of evolution were once dismissed as erroneous and had to win through over bitter opposition. Chinese history offers many similar examples.[48]

Information about new publications now comes from all quarters. The Chinese Academy of Social Sciences, for example, has announced that it will shortly resume publication of *Philosophical Studies* and *Economic Research* among others.[49] The provincial science and technology commission in Sichuan has announced that it will publish *Sichuan Science and Technology* which appeared early in 1978 after half a year's trial publication.[50]

Elzinga has pointed out there is now a very different atmosphere in that editorial boards have great freedom of expression, whereas earlier under the Gang of Four there was an attempt to enforce strong political "censorship" even over scientific journals, and the idea was to have party control over journals through party members filling editorial boards.[51]

Throughout the period since 1966 China has continued to publish technical books of many kinds even if the activity was at a low ebb in late 1960s. In more recent years approximately 2000 titles, new ones or reprints, covering the various fields of science and technology, have been published every year.[52] However, there can be no doubt the Chinese system for information and publication suffered, in particular from the poor coverage of foreign developments which was restricted to relatively few high-level research institutes.

The question of the role of the individual versus the collective in science and technology has been a hotly debated topic throughout the Cultural Revolution. There can be no doubt that groups and collective work is essential for any large research project. This can be seen in the organization of such projects in advanced capitalist countries. This should also be reflected in publishing, it was argued during the Cultural Revolution; individuals should not use articles in scientific journals and the publication of books to boost their ego and enhance their status among colleagues.

Today one can see a clear tendency towards a return of individual authorship for scientific articles; one reason being that it should be possible to question individuals on

[48] *On the Correct Handling of Contradictions Among the People,* Beijing, 1966, revised translation, Foreign Languages Press, pp. 35–36.

[49] NCNA in English, Dec. 25, 1977; BBC FE/5702/B11/18.

[50] Sichuan provincial service, Nov. 27, 1977; BBC FE/5695/B11/17.

[51] A. Elzinga, "Red and Expert", mimeograph, 3 parts, Institute for Theory of Science, University of Gothenburg, 1977–8.

[52] E. Baark, R. Jonsen, and D. Wagner, *Survey of PRC Literature on Science and Technology. A Bibliography, Partly Annotated,* Research Policy Programme, 1977, 46 pp.

specific aspects on the development of frontier science and technology which such articles are supposed to cover. However, book publishing still seems to be dominated by collectives, as is evident from a survey of 330 books in electronics, metallurgy, and agricultural machinery.[53] Table 6 shows that in 1970, 1971, and 1972 every book in the survey was written by some kind of a collective. In 1973 came a rapid re-emergence of individual authors who were responsible for more than one-third of the titles; this ratio dropped considerably in the later years (1974—7) to around 20%. Erik Baark points out that the recent low percentage is surprising considering the new policy concerning intellectuals and research initiated after the fall of the Gang of Four. However, there are plausible explanations. First, the effects of Chinese policy on these issues may be delayed. A continuation of the analysis in coming years will make it possible to answer this question. Second, the changes in policy concerning books in science and technology has not significantly affected the role of authors institutionalized since the Cultural Revolution. So the authorship of technical books in China is still predominantly a collective effort of either a group of individuals or a group of institutions.

The survey shows a distinct difference in the average number of books published in the two major categories of authorship: roughly 135,000 copies for collective authors against 68,000 for individuals. Baark draws the conclusion that the figures probably reflect the fact that collective authors usually write for a larger audience, i.e. more general works than do individual authors. This can possibly be interpreted as part of a Chinese policy on collective authorship and co-operation between factories and research institutes, so that the books which are intended for a large circle of readers would preferably be compiled as a collective effort by research institutes all over China. The more specialized areas with a smaller circle of readers could be provided with materials from individual authors. The survey also differentiates between various types of collectives and notes differences in the three sectors covered by the study (summarized in Table 7).

TABLE 6. *Individual versus Collective Authorship*

| Year | Individual (%) | Collective (%) |
|------|------|------|
| 1970 | 0.0 | 100.0 |
| 1971 | 0.0 | 100.0 |
| 1972 | 0.0 | 100.0 |
| 1973 | 38.5 | 61.5 |
| 1974 | 20.9 | 79.1 |
| 1975 | 18.8 | 81.2 |
| 1976 | 27.8 | 72.2 |
| 1977 | 16.7 | 83.3 |

Source: E. Baark, *Dissemination Structures for Technical Information in China – An Analysis of Three Industrial Sectors: Electronics, Metallurgy, and Agricultural Machinery,* interim report (Research Policy Programme), Lund, Aug. 1978.

[53] E. Baark, *Dissemination Structures for Technical Information in China – An Analysis of Three Industrial Sectors: Electronics, Metallurgy, and Agricultural Machinery,* interim report (Research Policy Programme), Lund, Aug. 1978.

TABLE 7. *Authorship in various sectors*

| Responsible "author" | Technological sector | | |
|---|---|---|---|
| | Electronic | Metallurgy | Agricultural machinery |
| Individual | 46 | 33 | 0 |
| Writing group | 18 | 48 | 3 |
| Research Institute | | | |
| University/"College" | 39 | 29 | 14 |
| Production unit | 20 | 12 | 25 |
| Publishing company | 15 | 0 | 0 |

Source: E. Baark, op. cit.

Baark points out that the ratio between individual and collective authors is approximately 1:2 for the electronics sector while it is approximately 1:3 for the metallurgical sector. Another difference between electronics and metallurgy is the predominance of research institutes in electronics. The obvious conclusion is that the electronics sector is more research orientated than the metallurgical sector. The figures for agricultural machinery may be somewhat misleading because the books purchased have a bias towards manuals and repair handbooks.

A new atmosphere was also noticeable in publishing when a national forum was convened by the State Publications Bureau of the State Council early in 1978. A report from the forum mentions that the publishing houses were called upon to put out more and better books and to pay special attention to editing, compiling, translating, and doing research on ancient classics and foreign works. Great efforts were expected in translating and publishing advanced theoretical and technical material on the natural science from other countries. The forum also declared that "efforts should be made to bring into full play the initiative of the professional writers as the backbone force . . . ".[54]

It may still require some time before the publication of academic and other journals in science is fully resumed, though rapid changes have already been made in the libraries. "Large numbers of Chinese and foreign books once banned by the Gang of Four have now been put back on the public shelves at Beijing Library" according to the Chinese news agency which adds:

> "All the books in the science and technology section are now available at the loan counter. Interested readers may pick up any of the 3000 kinds of periodicals, which report the latest scientific and technological advances in China and abroad, from the open shelves in the reading rooms.
>
> In the social sciences section, all books in philosophy, history, economics, law, and political science are available to the readers except for a certain amount reserved for the reference of departments concerned".[55]

Although we lack detailed information on the earlier situation, there can be no doubt that information handling and publication is being promoted with much fewer constraints than previously.

[54] National forum on publishing, NCNA, Jan. 12, 1978; BBC FE/5713/B11/1; and *People's Daily*, Dec. 20, 1977, BBC FE/5702/B11/18.
[55] Xinhua News Agency (Stockholm), 1978, No. 7 (Jan. 9).

# 5. Research Organization and Allocation of Resources

VARIOUS economic sectors in a society require different types and mixtures of R & D resources. So, the distribution of resources between basic research and grassroots application must also vary in the sectors. The level of development — compared with the industrialized countries or judged according to other criteria — and the subsequent need and availability of new resources, has consequences for the distribution of resources among the various parts of the innovation chain. This in turn has relevance for the forms utilized to organize R & D.

Science and technology in China covers a broad spectrum of activities. It includes big science projects like satellites, nuclear weapons, high-energy research, advanced electronics, as well as the diffusion of industrial innovations. It also includes the technology needed in big and small enterprises as well as the popularization of any scientific and technical knowledge required in agriculture or public health. Consequently, China has to find compromises under which resources can be allocated among the competing interests aside from overriding political conflicts.

### Research and Development Structure

Science and technology policy in China is formulated by the relevant bureau of the Communist Party, under the supervision of the Central Committee of the Chinese Communist Party and in co-operation with the State Planning Commission of the State Council. The State Commission of Science and Technology is then responsible for the overall planning, co-ordination, organization, and administration of the country's scientific and technological work. The various ministries, including those of Industry, Agriculture, and Public Health, operate research institutes within their own area of competence. This work is supervised and to some extent co-ordinated by the respective academies such as the Academy of Agriculture and Forestry Sciences. The administrative set-up at the national level is demonstrated in Fig. 2, which gives a simplistic presentation of some of the major central agencies responsible for science and technology.

Research activities in China are found in the following five sectors:

(1) institutes under the academies;
(2) universities and colleges;
(3) institutes under the ministries;
(4) provincial and municipally controlled research;
(5) Defence.

In all sectors the Commission has the responsibility to plan, co-ordinate, and administrate the science and technology efforts for the whole country.[1] In mid-1978 the Commission

---

[1] Fang Yi is president of the Commission which also has the following vice-presidents: Jiang Nanxiang; Wu Heng; Yu Guangyuan; Tong Dalin; Zhao Dongyuan; Jiang Ming.

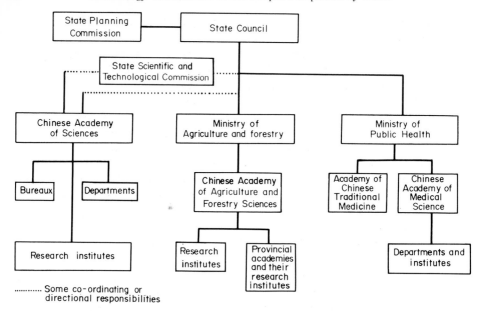

FIG. 2. *China's scientific and technological organizations.*
(From *Current Scene (Hong Kong)* Vol. XIV, No. 6.)

did not have more than 100 people although it is still expanding. It is organized into an administrative bureau, planning bureau, foreign affairs bureau, and various specialized bureaux. The State Science and Technology Commission has financing and co-ordinating functions although the main parts, e.g. applied industrial research, are financed and directly controlled by the ministries. However, the Commission is involved in all projects which cut across the ministries. The Commission deals with major questions such as establishing new institutes and co-ordinating large projects like the new centre for high-energy research; otherwise the lower level administrative units have been given the executive power.

During the Cultural Revolution, including the period up to the Fourth National People's Congress (NPC) in January 1975, a "Science and Education Group" under the State Council was the leading body besides the Academy of Sciences and the Science and Technology Association. "The Science and Education Group" at the time appears to have combined the function of the former Science and Technology Commission (which ceased to function in 1969–70), the Ministry of Higher Education, and the Ministry of Education. The Group was mainly concerned with general policy and in 1975 the organization seems to have disappeared after which the Academy of Sciences fulfilled a policy-making role, as it did in the early 1950s, until the reorganization of the Science and Technology Commission in September 1977.

Similar changes have also taken place at the provincial level. Shanghai Science and Technology Group changed its name to the Shanghai Commission for Science and Technology in November 1977. Aside from the name, more important changes are likely to be found in its responsibilities and in the composition of the leading body. The S & T Commission in Shanghai has five specialized bureaux including planning plus administrative offices and employs 100 people of which 50 are professionals. There is also a separate section, employing 200 people, which deals with the local research institutes belonging to the Academy of Sciences.

Research and development, including extension work in agriculture in China, are carried out at seven administrative levels. Organizations at four of these levels are operated by the State – central government, provinces, prefectures, and xian (county); the lower three levels are collectively operated – in the commune, the production brigade, and the production team (Fig. 3). The lower three levels include much if not most of the mass science and a considerable part of the regionalized R & D. Furthermore, the basic thrust of R & D in China after the beginning of the Cultural Revolution has been the expansion of the lower five of the seven levels mentioned. In view of this it is not surprising to find that a recent American delegation concluded that the Chinese have created what is probably the most effective system of agricultural extension to be found anywhere in the Third World. China's consistent efforts to redistribute opportunities for generating technological change by downplaying the role of professional elites and foreign technology may be particularly relevant within the traditional and rural sectors.

When reading the Chinese news one often gets the impression that since 1976 new administrative bodies for science and technology have been created at various administrative levels. This is apparently not the case. Instead, the names of the administrative units have been altered. Also, their membership and terms of reference in planning, supervising, and co-ordinating science and technology activities are likely to have undergone considerable changes.

A first big push for local-level science and technology came during the Great Leap Forward period in the late 1950s. Since then there have been varying degrees of national emphasis on such institutions, and there have been re-definitions of the role of local governments and economic institutions. However, the basic principle of a *network* of scientific and technological institutions at all administrative levels of society has not been lost.

By the early 1970s a R & D network was fairly well developed. This enables province-level initiatives in science policy, and in a sense justifies strong province-level leadership in many areas of science and technology work.

It will be useful, in briefly discussing these institutions, to see them in relation to the administrative structures of a province. The levels of administration and the types of scientific and technological institutions corresponding to these levels,[2] as well as the relations of the various units and their interaction, are presented in Fig. 3.

In the light of this it may be appropriate to view China's overall science policy in terms of a hierarchy of science policies, each with its own emphasis.

(1)  Nation – military R & D, energy, large modern industry.
(2)  Province – agricultural R & D, industrial development, medical R & D.
(3)  (a) Region – agricultural R & D, industrial adaptation.
     (b) Cities and large towns – industrial adaptations and applications.
(4)  County – agricultural adaptations, industrial transfer.
(5)  Commune – agricultural transfer, industrial transfer.

---

[2] This presentation appears in *Scientific Institutions in the People's Republic of China,* by R. P. Suttmeier, Seminar Paper DSTI/SPR/76.4, OECD Seminar on Science and Technology in the People's Republic of China, Paris, Jan. 20–23, 1976.

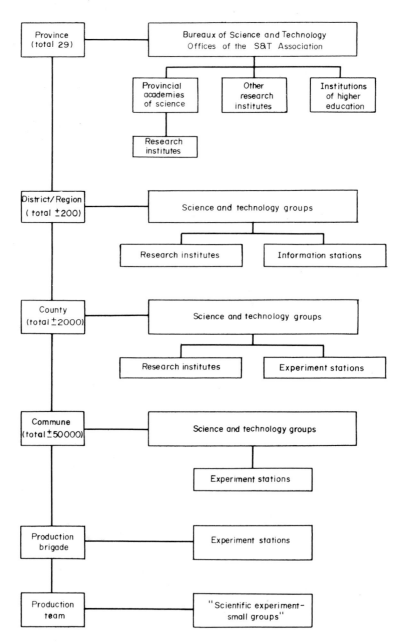

FIG. 3. *Science and technology institutions at various administrative levels.*
(From R. P. Suttmeier, *Science and Technology in the People's Republic of China,*
*OECD,* Paris 1977, p. 60, Chapter 3, on Scientific Institutions.)

The R & D system naturally poses a number of problems which can more easily be understood if we look at Fig. 4. The units depicted here cover a wide range of activities from break-through innovations in modern industry and defence to small incremental and adaptative innovations in the traditional, basically rural, sector of the economy.[3]

The modern sector is organized in large or very large units. This also applies to the R & D institutes which usually have a staff from several hundred to more than one thousand. Most of the industrial enterprises requiring R & D results are medium to large with at least a few hundred up to many thousand employees. Similarly, the institutions of higher learning have a potential enrolment of 10,000 or more. Consequently, the efficiency of the vertical transfer mechanisms has not always been satisfactory – a situation which is similar to that in the USSR.

## The Chinese Academy of Sciences

The Chinese Academy of Sciences (Academia Sinica) is the most important centre for scientific research and has at times had a semi-autonomous position under the government. Other research institutes are found within central and provincial government ministries and departments, within industrial enterprises and companies and at the universities. Following the development of regional science and the mass-based agricultural scientific

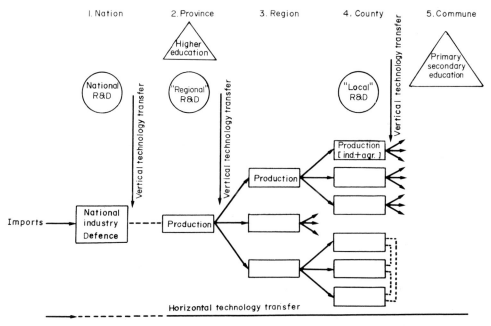

FIG. 4. *Technology transfer linkages in China*
(From J. Sigurdson, *Transfer of Technology to the Rural and Collective Sectors in China, in Science and Technology in the People's Republic of China,* OECD, Paris 1977, p. 179.)

[3] Obviously it is essential to differentiate between *vertical technology transfer* and *horizontal technology transfer.* Vertical technology transfer takes place when information is transmitted from basic research to applied research, from applied research to development, and from development to production. Horizontal technology transfer takes place when technology used in one place, organization, or context is transferred and used in another place, organization, or context.

network, research units, although of a less-sophisticated nature, can also be found within the relevant units of the lower administrative levels, i.e. prefectures and counties.

The Chinese Academy of Sciences, or Academia Sinica as it is sometimes called, has a rather special position. This is derived from the fact that it runs its own institutes which are mainly in basic and theoretical sciences and that it is today at least partly shouldering the responsibility for postgraduate training, in collaboration with the Ministry of Education. The Cultural Revolution influenced the Academy quite considerably, and one of the immediate effects was the dismantling or transfer of a large number of units to the lower levels of the Chinese administrative hierarchy. Changes are now being made which may return the Academy to the situation before the Cultural Revolution. The change in the number of institutions belonging to the Academy at various times since the mid-1960s can be seen in Table 8.[4]

In spring 1978 the number of institutes under the Academy had increased to more than 70, and foreign visitors were a few months later told that the number was around 90. Table 9 lists 72 institutes which according to official information in June 1978 belonged to the Academy. Appendix IV also mentions the major research areas of the various institutes.

Of the research institutes listed, 30 are located in Beijing and 11 in Shanghai. The third next important research centre is Lanzhou in Kansu Province with 5 institutes. So, obviously Beijing and Shanghai are the major research centres in the basic and theoretical work for which the Academy of Sciences is responsible. This is also true for many of the applied sciences of which electronics is one example. In order not to produce too great a geographical concentration, it has been decided that new institutes should not be located in Beijing.

The Academy of Science is presently organized around four departments or bureaus (since the social sciences have been formed into a separate Academy):

(1) Mathematics, Physics, and Chemistry.     (3) Biosciences.
(2) Geosciences.     (4) New Techniques.

TABLE 8. *Chinese Academy of Sciences: number of institutes under the different departments*

| Year | Physics, mathematics, & chemistry | Bio-sciences | Geo-sciences | New tech-niques | Philosophy and social sciences | Total |
|---|---|---|---|---|---|---|
| 1964 | 22 | 23 | 15 | 24 | 12 | 96 |
| 1966 | 22 | 23 | 15 | 24 | – | |
| 1967–8 | 14 | 28 | 15 | 18 | 14 | 89 |
| | | | | | | (+18 branches) |
| 1975 | | | | | | |
| Direct control | 9 | 3 | 5 | 1 | | |
| Jointly with ministry of province | 6 | 5 | 2 | 5 | | |
| Total | 15 | 8 | 7 | 6 | – | 36 |
| 1978 (March) | 24 | 17 | 17 | 14 | 0 | 72 |

[4] B. Billgren and J. Sigurdson, *An Estimate of Research and Development Expenditures in the People's Republic of China in 1973*, OECD Development Centre, Paris, July 1977.

TABLE 9. *The Chinese Academy of Sciences – institutes (June 1978)*

MATHEMATICS, PHYSICS, AND CHEMISTRY
1. Institute of Mathematics (Beijing)
2. Institute of Mechanics (Beijing)
3. Hubei Institute of Rock and Soil Mechanics (Wuhan)
4. Institute of Physics (Beijing)
5. Institute of Atomic Energy (Beijing)
6. Institute of High Energy Physics (Beijing)
7. Yunnan Branch Institute of Cosmic Rays (Kunming)
8. Institute of Modern Physics (Lanzhou, Gansu)
9. Shanghai Institute of Nuclear Physics
10. Institute of Acoustics (Beijing)
11. Jilin Institute of Physics (Changchun)
12. Xinan Institute of Physics (Leshan, Sichuan)
13. Institute of Plasma Physics (Hefei, Anhui)
14. Institute of Chemistry (Beijing)
15. Institute of Photochemistry (Beijing)
16. Institute of Environmental Chemistry (Beijing)
17. Institute of Chemical Metallurgy (Beijing)
18. Shanghai Institute of Organic Chemistry
19. Shanghai Institute of Silicate Chemistry and Engineering
20. Fujian Institute of Material Structure (Foochow)
21. Lanzhou (Gansu) Institute of Chemical Physics
22. Dairen (Jilin) Institute of Chemical Physics
23. Jilin Institute of Applied Chemistry (Changchun)
24. Guangdong Institute of Chemistry (Guangzhou)

GEOSCIENCES
25. Beijing Astronomical Observatory
26. Purple Mountain Observatory (Nanjing)
27. Shanghai Observatory
28. Yunnan Observatory (Kunming)
29a. Institute of Geography (Beijing)
29b. Institute of Geography Section 2 (Beijing)
30. Institute of Geology (Beijing)
31. Institute of Geophysics (Beijing)
32. Institute of Atmospheric Physics (Beijing)
33. Lanzhou (Gansu) Institute of Plateau Atmospheric Physics
34. Guiyang (Guizhou) Institute of Geochemistry
35. Institute of Vertebrate Palaeontology and Palaeanthropology (Beijing)
36. Nanjing Institute of Geopalaeontology
37. Lanzhou (Gansu) Institute of Glaciers and Frozen Soil
38. Lanzhou (Gansu) Institute of Deserts
39. Institute of Oceanology (Qingdao, Shandong)
40. South China Sea Institute of Oceanology (Guangzhou)
41. Committee for General Investigation of Natural Resources (Beijing)

BIOSCIENCES
42. Institute of Zoology (Beijing)
43. Yunnan Institute of Zoology (Kunming)
44. Shanghai Institute of Entomology
45. Institute of Microbiology (Beijing)
46. Institute of Psychology (Beijing)
47. Shanghai Institute of Physiology
48. Institute of Biophysics (Beijing)
49. Shanghai Institute of Biochemistry
50. Shanghai Institute of Cell Biology
51. Shanghai Institute of Materia Medica
52. Institute of Genetics (Beijing)
53. Institute of Botany (Beijing)
54. Kwangtung Institute of Botany (Guangzhou)

*(continued overleaf)*

55.  Yunnan Institute of Botany (Kunming)
56.  Shanghai Institute of Plant Physiology
57.  Nanjing Institute of Pedology
58.  Institute of Forestry and Pedology (Shenyang, Liaoning)

NEW TECHNIQUES
59.  Institute of Computing Technology (Beijing)
60.  Shenyang (Liaoning) Institute of Computer Technology
61.  Institute of Automation (Beijing)
62.  Shenyang (Liaoning) Institute of Automation
63.  Institute of Electrical Engineering (Beijing)
64.  Institute of Electronics (Beijing)
65.  Shanghai Institute of Technical Physics
66.  Changchun (Jilin) Institute of Optical and Precision Instruments
67.  Shanghai Institute of Optical and Precision Instruments
68.  Sian (Shaanxi) Institute of Optical and Precision Instruments
69.  Institute of Semiconductors (Beijing)
70.  Shanghai Institute of Metallurgy
71.  Institute of Metals (Shenyang, Liaoning)
72.  Institute of History of Natural Sciences (Beijing)

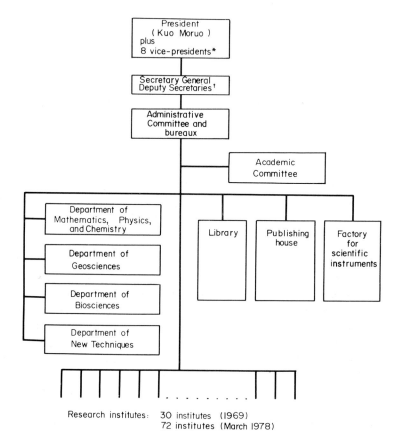

FIG. 5. *The Chinese Academy of Sciences organizational outline*
*(March 1978)*
*Vice-presidents:  Fang Yi, Li Zhang, Zhou Peiyuan, Tong Dizhou, Hu Keshi,
Qian Sanqiang, Hua Legeng, Yen Jizi.
†Deputy secretary generals: Jin Lisheng, Yu Wen, Qian Sanjiang, Hao Mengbi,
Gan Zhongdou.

The Academy is responsible for a major national scientific and technical library located in Beijing, runs a substantial publishing company and is responsible for the operation of a scientific instrument factory. The organization as perceived by a limited number of foreign visitors is given in Fig. 5.

The significance of Shanghai as a research centre is clear from the fact that the total staff of all its research institutes, including industrial research, numbered 60,000 in 1978. Of the 800,000 qualified researchers set as a target for 1985 by Fang Yi at the National Science Conference, planners in Shanghai expect to have more than 10%.

Nunn[5] mentions that a typical research institute within the Chinese Academy of Sciences or government consists of one or more specialized laboratory sections or departments and subsidiary units such as factories. Collaborative relationships between specified universities or factories are usually governmentally directed even if the period of open-door research may partly have changed this situation. Specific projects involve co-operation between research institutes. Today there is usually one director, who may be a scientist or administrator with close party affiliation, and one or more deputy directors. From 1966 until quite recently the institutes were usually under the control of the revolutionary committee, with participation of workers and political cadres as well. Employees number from 300 to 1000, including a small number of senior scientists and a large number of intermediate and younger scientists.

The size of some institutes under the Academy, and those most frequently visited by foreigners, is given in Table 10. Some of the figures are occasionally contradictory, and the interested readers should consult the original report which provides detailed information on the sources utilized.

## Research Resources

The total number of research institutes is 2,400 above county-level; including low-level agriculture-related institutes the total is 6,000. A recent tabulation made by Susan Swannack Nunn on 1,000 identified research institutes showed 43% in life sciences, 34% in engineering and technical sciences, and 22% in physical sciences.[5]

We have attempted to provide a breakdown of the R & D budget by major categories in Table 11. This shows that basic research receives 4% of the total R & D budget. The figure is possibly on the high side because it includes all R & D activities carried out within the institutes either under the direct control of the Academy of Sciences or under dual jurisdiction of the Academy and another agency. On the other hand, basic research may also be carried out in other sectors. The estimates used here can be challenged on many counts but may still be useful in the absence of official data from China.

Even the Chinese planners may not at the time of the National Science Conference have had any access to any detailed statistical information on the number of researchers, at least not in the natural sciences. This is obvious from the announcement that the State Planning Commission, the State Scientific and Technological Commission, the Ministry of Civil Affairs, and the State Statistical Bureau agreed to conduct a nationwide survey of scientific and technical personnel in the field of natural sciences. This was said to be

[5] S. S. Nunn, "Research institutes in the People's Republic of China", *US–China Business Review,* Mar. – Apr. 1976.

TABLE 10. *Seventeen research institutes in Beijing under the direct control of the Chinese Academy of sciences**

| Institute | Total staff | | | | | Engineers and scientists | Techni-cians |
|---|---|---|---|---|---|---|---|
| | 1972 | 1973 | 1974 | 1975 | 1978 | 1973 | 1973 |
| Atmospheric Physics | – | – | – | – | – | – | – |
| Atomic Energy | – | 1000 | 1000 | – | 1500 | 400 | 300 |
| Biophysics | – | 220 | 220 | 400 | – | 110 | 40 |
| Chemistry | 1100 | 600 | 600 | 600 | – | 300 | (150) |
| Computing Technology | 1000 | – | – | – | – | – | – |
| Genetics | – | 350 | 400 | 200 | – | 200 | – |
| Geography | – | – | – | – | – | – | – |
| Geology | – | – | 1000 | – | – | 400 | – |
| Geophysics | (300) | 350 | – | – | – | (200–300) | – |
| High-energy Physics | – | – | – | – | – | – | – |
| Mathematics | – | – | 300 | – | – | (200) | (40) |
| Microbiology | (400) | – | 400 | 400 | – | (255) | – |
| Metrology | – | – | – | – | – | – | – |
| Physics | 600–700 | 600–700 | 600–700 | 400 | – | 500 | – |
| Beijing Observatory | 200 | – | 300 | – | – | – | – |
| Semiconductors | 900 | – | – | 900 | – | (500) | – |
| Vertebrate Paleontology and Paleoanthropology | – | – | – | – | – | – | – |

*B. Billgren and J. Sigurdson, *An Estimate of Research and Development Expenditures in the People's Republic of China in 1973,* OECD Development Centre, Paris, July 1977.

TABLE 11. *Breakdown of R & D expenditures by major categories (1973 estimate)*

| | Manpower | | | | Expenditure | | | |
|---|---|---|---|---|---|---|---|---|
| | Thousands | | Percentage | | Billion yuan | | Percentage | |
| | I* | II† | I | II | I | II | I | II |
| Basic Research | 21.3 | 21.3 | 2 | 4 | 0.083 | 0.083 | 3 | 4 |
| Agriculture and natural resources excluding energy | 392 | 102 | 45 | 17 | 0.601 | 0.311 | 24 | 14 |
| Medicine and public health | 102 | 102 | 12 | 17 | 0.311 | 0.311 | 13 | 14 |
| Defence | 100 | 100 | 11 | 17 | 0.517 | 0.517 | 21 | 24 |
| Manufacturing, energy, and transportation | 261 | 261 | 30 | 45 | 0.970 | 0.970 | 39 | 44 |
| Total | 876 | 586 | 100 | 100 | 2.48 | 2.19 | 100 | 100 |

\*   I, mass scientific network in agriculture included.
†   II, mass scientific network in agriculture excluded.

PLATE 9. China is in the top league of the world's fishing nations and is the leader both in pond and marine fish culture. The annual catch of cultured fish is over 3 million tons. Researchers in Guangdong Province have introduced new strains and Guangdong Provincial Aquatic Products Research Institute is the lead agency. (Chung Lin, section chief for freshwater research, studies the embryonic development of fish.)

"warranted in order to gain a comprehensive understanding of their numbers, levels of schooling, distribution, and employment . . . "[6]

Research and development in agriculture and natural resources received, according to available estimates, 14% of the total R & D resources. The figure does not quite indicate the important role given to agricultural research since the sector is expanding and accounts for a larger share of manpower than the figure indicates. If we include the mass scientific network in agriculture − discussed at some length below − the share for agriculture may be as high as 24%. Medicine and public health receives 14% of R & D resources.

The 24% figure for defence, which may be much too low or much too high, should still be seen as an indication of the importance of military R & D in China. It is natural to find that manufacturing, energy, and transportation receive a very considerable part of the total resources, a reflection of their importance along with agriculture.

These sectors add up to a total R & D budget of 2.48 billion yuan in 1973, which approximately equals $US 1.25 billion. This is roughly 0.6% of the gross national product of the People's Republic of China in 1973 (in 1973 $US). This would indicate, if the

[6] *Nationwide Survey of Scientific and Technical Personnel*, NCNA June 23, 1978; BBC FE/5855/ B11/15.

estimated budget is a close approximation, that the general assumptions about the relative level of R & D expenditure in the People's Republic of China have generally been on the high side.[7]

The estimates made here do not include the humanities and social sciences. The reason is twofold. The humanities receive very limited funds compared with the other major categories which are dealt with here. Second, most social science research is carried out as investigations under the direct control of party organs, which sets social sciences aside from all other R & D activities discussed in this book.

Founding procedure can be demonstrated by the plans for a tandem accelerator to be set up at the Atomic Energy Institute (AEI) outside Beijing.[8] The project will involve collaboration with the Institute of High Energy Physics, also in Beijing, and part of AEI until 1973. Total costs are calculated to reach 10 million yuan, and the project is expected to run 3–5 years. A project of this magnitude involved not only the institutes themselves and the Academy of Sciences, but also the Science and Technology Commission before the final decision was taken in the State Council (Fig. 6).

We have also attempted to use our estimate of the R & D budget to indicate the amounts that go to the science and technology units at the lower levels. Here we have included all units at provincial and lower levels which are specially geared to local and rural needs. The amounts received may be approaching 20% of the total national budget for science and technology and is still more in manpower terms (Table 12).

Basically, the task in the provinces and at the still lower administrative levels is not so much one of undertaking new research projects as that of extension. Research carried out at these levels should be of immediate relevance. So the role of the lower levels is more to complement than substitute the R & D which is under the responsibility of the higher levels.

The various units included in regional science as well as the organizational set-up are given in Tables 10 and 11 in the chapter on mass science. The information below gives an indication of the mass science network as it existed in parts of the Hebei province during the author's visits in 1971 and 1973.

---

[7] Against the repeated emphasis on the gaps in science and technology between China and advanced countries and with the evidence indicating that this sector of the society was badly neglected or interfered with for a number of years, it appears that earlier estimated R & D figures are higher than what they actually were. This has been pointed out by Leo Orleans and is also supported by information and impressions gained during a recent visit to China by the author (B. Billgren and J. Sigurdson, *An Estimate of Research and Development Expenditures in the People's Republic of China in 1973*, OECD Development Centre, Paris, July 1977).

Earlier estimates of the People's Republic of China R & D manpower and expenditure have consequently been revised downwards. The reasons for the earlier estimates being too high can basically be traced to the following two reasons. First, the belief that R & D spending has followed the expansion of the economy – in particular the relative rapid growth of industrial output and maintained at a fixed ratio of 1% or higher of GNP – is apparently completely false. Second, the assumption that China developed all science and technology sectors along ambitious plans set up in the 1950s does not conform with reality. Furthermore, accepting the 1977 estimates by Sigurdson and Billgren and comparing them with figures of other countries, e.g. Japan and India, yields untenable ratios. Consequently it has become necessary to revise downwards the earlier estimate, new ones now being presented in Tables 11 and 12.

[8] The Atomic Energy Institute presently (Mar. 1978) has a total staff of 1500 people of which 400 are university graduates – between 100 and 200 have a doctorate. The annual research budget is 3 million yuan, which apparently does not include staff wages and salaries at a total cost of around 2 million yuan.

Zunhua County has 43 people's communes, all of which have technology populariza-tion stations. Aside from commune members, each has at least one agricultural technician who has been sent down from a higher-level organization. These commune stations generally have experimental fields and are usually responsible for all aspects of agricultural technology and veterinary technology, as well as farm machinery improvements.

The brigades have agricultural technology groups with from 5 to 13 people. These are *three-in-one* groups (combining leadership personnel, technicians, and peasants), which have at least one cadre from the brigade revolutionary committee (usually the chairman), at least one agricultural technician, and for the rest, veteran peasants and educated youth. In Zunhua more than 390 brigades out of a total of 692 have experimental farms, which are under the leadership of the brigade agricultural technology groups. Finally, the teams have agricultural technology groups with from three to seven members.

If, then, the agricultural technology people at commune level, the members of public health technology groups and of technology groups in industrial enterprises are added together, the total should be roughly 25,000 according to information provided by the

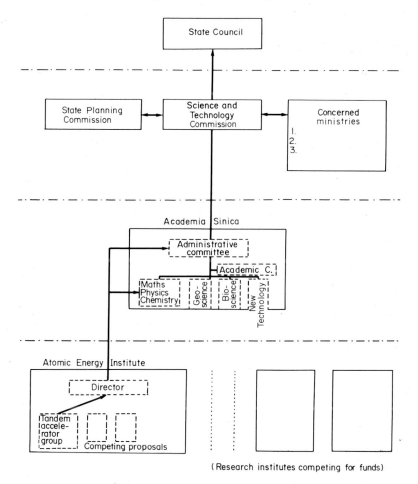

(Research institutes competing for funds)

FIG. 6. *Decision levels for funding large research projects ("tandem accelerator" at the Atomic Energy Research Institute of the Academia Sinica)*

TABLE 12. *Estimated R & D expenditure and manpower in 1973*[a]

| | Manpower | | | | Expenditure | | | |
|---|---|---|---|---|---|---|---|---|
| | Thousands | | Percentage | | Billion yuan | | Percentage | |
| | I* | II† | I | II | I | II | I | II |
| A. Total R & D costs in the Academy of Sciences, institutions of higher learning and industrial research (assuming that the wage bill is 40% except for defence sector 30%) | 349 | 349 | 40 | 59 | 1.48 | 1.48 | 60 | 68 |
| B. R & D in life sciences, agriculture, and medicine (20% of categories in A) | 70 | 70 | 8 | 12 | 0.27 | 0.27 | 11 | 12 |
| C. Mass scientific network (7 million people, 2 hours per week, each man-year valued at 1000 yuan) | 290 | – | 33 | – | 0.29 | – | 12 | – |
| D. Regional science | 168 | 168 | 19 | 29 | 0.44 | 0.44 | 18 | 20 |
| Total | 877 | 587 | 100 | 100 | 2.48 | 2.19 | 101 | 100 |

\*  I, mass scientific network in agriculture included.
†  II, mass scientific network in agriculture excluded.

[a]For a discussion on methodology the interested reader should consult Billgren and Sigurdson, op. cit.

Zunhua County Science and Technology Group. This would mean that slightly more than 4% of the total population of roughly 600,000 in the county has been enrolled in the mass scientific network. The equivalent figure for the whole of Hebei Province is 1 million, which is about 2% of a population of 48 million.

These two subsectors of the Chinese science and technology system may together include almost 40% of the manpower utilized. In financial terms it is less than 20%. The figures relate to 1973, and both regional science and the mass scientific network are likely to have increased their share of the total national resources. However, the recent policy directives clearly indicate that the R & D sectors serving the modern economy will again be emphasized — while other needs are not neglected. So mass science and regional science may keep their relative positions.

### The science and technology commissions

The provincial commissions on science and technology were in their present role re-established in 1977—78 shortly after the State Science and Technology Commission (SSTC) was re-established in autumn 1977. All commissions, both SSTC and the regional ones, have a similar organizational structure which is exemplified in Fig. 7. However, the commission in Shanghai as well as the national one has a more elaborate bureau structure in order to handle its more diversified responsibilities. Generally, the commissions have an administrative office in addition to bureaus for planning, policy research, results, equipment and international relations. Several regional commissions have a staff of around 100 while SSTC has more than 200 in late 1979. In Shanghai a considerable amount of work is done by external groups attached to the commission.

There is a distinct division of responsibility between the regional commissions and the State Commission. The former each have a budget which is at most 5 per cent of SSTC's and are only responsible for research of local nature while SSTC controls all research of national and comprehensive nature — in cooperation with the Academy of Sciences.

The tasks of the State Science and Technology Commission are reported to be the following:

1. Carry out the country's policy for scientific and technological development.
2. Formulate long term and annual plans.
3. Coordinate ministries and provinces/municipalities for import projects.
4. Evaluate important research results and diffuse them into production.

FIG. 7. *Science and technology commissions: Bureau structure*

| State | Guangdong | Shanghai | Jiangsu | Tianjin |
|---|---|---|---|---|
| Office | Office | Office | Office | Office |
| Bureau 1<br>— Basic research<br>— Theoretical research | Agriculture | Bureau 1<br>— New material<br>— Metallurgy<br>— Chemical, Light and Textile Industry | Agriculture | Industry |
| Bureau 2<br>— Energy, Resources<br>— Materials | Industry | Bureau 2<br>— New technologies<br>— Electricity, Communications<br>— Transportation, Energy | Industry | Non-industry |
| Bureau 3<br>— Transportation<br>— Industry | . . . . . . . . ? | Bureau 3<br>— Basic Research<br>— Agriculture<br>— Medicine and Hygiene<br>— Finance and Trade | | Agriculture |
| Bureau 4<br>— Agriculture, Health<br>— Light Industry | | | | |
| Bureau 5<br>— Peaceful Use of Nuclear energy | | | | |
| Planning<br>Policy Research<br>Scientific Results<br>Research Facilities<br>Foreign Affairs | Planning<br><br>Scientific Results<br>Research Facilities<br>Foreign Affairs | Planning<br>Policy Research<br>Scientific Results<br>Research Facilities<br>Foreign Affairs<br>Organization<br>Propaganda | Planning<br>Policy Research<br><br>Research Facilities<br><br>Political | Planning<br>Policy Research<br><br><br>Foreign Affairs |

5. Train scientific and technical personnel and take measures for retraining.
6. Distribute funds for science and technology activities and provide necessary equipment.
7. Support international relationships and collaboration in science and technology.
8. Formulate the necessary regulations, laws and other legal specifications in the area of science and technology — to be approved by the National People's Congress.

SSTC receives proposals and suggestions from the lower levels — ministries, provinces and the three cities with provincial status, Beijing, Shanghai and Tianjin. SSTC then summarizes, and again discusses with the lower levels before finalizing the plans. The document for the plan is then handed over to the State Council for final approval. The annual plan may have 400–500 projects which are all formulated in view of the require-ments in the long-term plan.

Most projects are screened by a number of experts — scientists, engineers and professors who are organized in a number of committees some of which are sent to the localities to obtain a fuller understanding. At present SSTC has more than 60 such committees and works in close cooperation with the National Association for Science and Technology which organizes all professionals all over China.

Researchers working in the research institutions — i.e. above county level which basi-cally means academy institutes, ministerial institutes and universities — amounts to 310,000, according to Chinese official statistics, in more than 2,400 institutes. Including the agriculture-related institutes at county level and below the total number of institutes is 6,000. With the researchers at the lower, mainly collectively controlled research insti-tutes in rural areas, the total is reported to be 420,000 which amounts to 7 per cent of a total of 6 million scientific and technical manpower in the 1978 national survey. See figures below for relative shares of China's total technical and scientific manpower.

| | | |
|---|---|---|
| Engineering | 36% | |
| Medicine and health | 30% | |
| Training and education | 20% | |
| Research | 7% | 420,000 (includes collective research) |
| Agriculture | 7% | |
| Total | 100% | 6,000,000 |

The recently released figure for national R & D gives a total budget for 1979 of 5.87 billion yuan which is said to be an increase of 10 per cent over 1978, which we interpret to be a considerable increase over the early 1970s. This figure includes invest-ment in facilities and laboratories, wages and all expenses for general use as well as costs for the specific projects including research in pilot plants. It is said to constitute almost exactly one per cent of the total production value in agriculture and industry and includes research in medicine and social sciences as well.

A major change can be seen in the resources that go to basic research since the Academia Sinica was restored to its previous functions in the mid- and late 1970s. The total budget of the Academia Sinica is today in the region of 1.050 million yuan, i.e. 18 per cent of the total national budget for R & D in 1979. A little more than one half or 585 million yuan goes to basic research — according to the State Science and Technology Commission — most of which is carried out within the Chinese Academy of Sciences. The relative share for applied research and development are given below.

| | | |
|---|---|---|
| Basic research | 10% | 585 |
| Applied research | 20% | 1,175 |
| Development | 70% | 4,110 |
| Total | 100% | Y 5,870 million |

The total national budget of 5.87 billion yuan is split into almost equally large por-tions. One half is controlled by the State Science and Technology Commission which finances the Academy of Sciences and all major national projects which cuts across the national ministries. The other half includes the science and technology budget of provinces which receive approximately 850 million yuan most of which goes to a limited number of provinces with the rest of that half going to the ministries which also include defence research. See Fig. 8.

FIG. 8. *China's R & D resources in 1979 (million yuan)*

| | Provincial | | | | National | |
|---|---|---|---|---|---|---|
| | Guangdong | Shanghai | Jiangsu | Tianjin | Ministry | Commission |
| Regional S & T Commissions | | 130 | 100 | = 850 | | |
| Ministry research including defence | | | | | 2,050 | |
| | | | | Subtotal 1: | 2,900 | |
| Academy of Sciences | | | | | | 1,050 |
| Major projects + other activities controlled by the state science and technology commission | | | | | | 1,850 |
| | | | | Subtotal 2: | | 2,900 |
| | | | | TOTAL | | 5,800 million |

# 6. Education and Training

CHINA, like many developing countries, is facing an educational dilemma. On the one hand, it is necessary to ensure that the benefits of education reach down to the broad masses of the people. On the other hand, it is equally important for a country aiming at rapid industrialization to have a wide range of highly qualified professional cadres. These two different needs, equally urgent, have not been reconciled. The consequences among other things appear to have been that the training of specialists and the university system were neglected in the past. Thus, institutions for training of engineers and scientists will now have to expand considerably if the desired high rate of growth in industrial and agricultural production is to be realized.

Another major issue is the still lagging universal education and the low rate of literacy in higher age groups. This makes it necessary to rely extensively on non-formal education in order to supplement the normal programmes but also to reduce costs for specialized education. A primary education oriented towards practical subjects is necessary in order to develop the rural areas where the majority of the population is still to be found. This will help to support programmes for increased productivity in agriculture and in order to promote equality.

### National Education Conference and National Education Programme

The National Education Programme 1978–85 has the following two basic concepts:[1]

(1) Walking on two legs and running many types of schools;
(2) Raising the quality of education.

At the national conference the Minister of Education pointed out that it is necessary to have various schools for higher education. With an emphasis on the four-year or five-year regular universities, vigorous efforts will also be made to develop two-year or three-year specialized institutes and to set up, on a trial basis, two-year or three-year basic college classes or basic colleges. New colleges and universities will be established more rapidly so that each province will basically be able to meet its own needs for trained personnel in general, including those in the fields of liberal arts, science, engineering, agriculture, medicine, and the training of teachers.

In addition China will develop television, radio, correspondence courses, night school, and other forms of spare-time education, and also develop spare-time education in factories, mines and other enterprises, and research and planning organizations, so that the large numbers of people settling in the countryside can take advantage of them and reach the levels of secondary vocational schools or universities. In developing television and radio courses a start will be made with projects which are less expensive and which produce quick and large-scale results.

[1] Education Minister's Report to the National Education Conference (NCNA, June 10, 1978); BBC FE/5843/B11/1.

Efforts are also being made to investigate, reorganize, and improve the regular senior middle schools. All kinds of agricultural middle schools, secondary vocational schools, and technical schools, most of which are part work and part study schools, will be developed in order to greatly increase their proportion in the entire secondary education system.

With teachers obviously being a critical resource, experts and scholars from the Academy of Sciences and the various ministries and departments can expect to become part-time teachers; institutions of higher education should exchange teachers among themselves. The key primary and secondary schools can select outstanding teachers from among the teachers of other schools and various departments while encouraging college teachers and scientists and technicians to teach part time. Furthermore, it has been considered necessary to restore and establish regular competence tests for teachers and a system of ranking and promotion in order to encourage good work. The ranking and promotion of teachers will be done in accordance with their actual teaching abilities and academic level and not on the basis of seniority. The minister said that teachers with outstanding performance can skip promotion ranks irrespective of their academic standing or seniority. Particularly outstanding middle-school and primary-school teachers may also be given a special grade.

The concept of priority, or key, schools is considered necessary for improving quality. Obviously the requirements from all schools cannot be adequately met at present due to the shortage of teachers, of teaching aids and funds. So a number of key schools have been started again at primary, secondary, and tertiary level. These schools now enrol students with better educational foundations, and themselves have better facilities. Within ordinary schools, key classes and areas of specialization will be decided upon. Key schools, specialty areas, and key classes are today said to be required in order to turn out better students and to help improve the quality of education as a whole.

There are consequences throughout the system, and the institutions of higher education are expected to give priority to those from the key schools during the unified college student enrolment. The areas from which the key primary and secondary schools may recruit students should be enlarged by removing the restriction that students should go to the nearest school.

Rural branch schools are being closed down by the Guangzhou Municipality; they put a great strain on the teachers.[2] The problem of lagging education in rural areas is obviously recognized in a number of places:

> Conditions are of course still better in city schools than in rural schools, and therefore children of workers, cadres and intellectuals are likely to be more competent in college entrance examinations for the time being . . . the way to narrow the gap is to raise the level of primary and secondary education in the rural areas.[3]

The changed emphasis on manual labour is related to the quality of education and the students are now mainly involved in study. The period of physical labour for pupils above third grade is fixed for the time being at 4 weeks in each year, that for middle school students at 6—8 weeks, and that for college students at 4 weeks. College students generally do physical labour that is closely related to their specialized training.

In the following a choice has been made to focus the discussion on two aspects —

[2] Guangzhou abolishes rural branch schools (Guangzhou, July 29, 1978); BBC FE/5883/B11/7.
[3] College enrolment statistics in a Sichuan County (NCNA, Aug. 3, 1978); BBC FE/5883/B11/9.

PLATE 10. Research workers using a portable, electrostatic, X-ray unit to examine virus pneumonia at the Research Institute of Agricultural Science in Ningxia Province.

university education and postgraduate studies — which we consider more related to the development of science and technology and to deal only summarily with primary and secondary education which, of course, is of basic importance for China's development strategy. However, China's dual sector development will require great changes both in the rural areas and in the modern industrial sector, and new educational needs will have to be met through developing both primary and university education.

### Primary and Secondary Education

A draft programme for a tentative implementation of the full-day, ten-year primary and middle school teaching system:

> stipulates the tasks, schooling system and basic principle for the formulation of teaching plans for primary and middle schools, the time for "principal studies" and "subsidiary studies" and the complete range of activities and institution of courses in these schools. Full-day primary and middle level schooling calls for 10 years of schooling — five for primary school and five for middle school. Of the five years at the middle school level, three are for junior classes and two for senior classes. All students will begin the school year in the autumn, in those localities where conditions permit, enrolment of children at the age of six or six and-a-half may be practised step by step.[4]

However, in education China will pursue a two-leg policy and simultaneously strive for "popularization and higher standards". This means that there will be full-day schools,

[4] Education Ministry's Programme for full-day, ten-year schooling (NCNA, Feb. 12, 1978); BBC FE/5742/B11/17.

work-study schools, and spare-time schools as well. Most of the primary and middle schools in the rural areas are still practising the nine-year school system which will gradually be turned into full-day ten-year schooling. Until then the teaching programmes may be formulated by the respective provinces, municipalities, etc. However, full-day schools with a ten-year programme must improve school management and have "the top three leaders installed as soon as possible".[4]

An attempt will be made here to indicate the· effect of the expansion of primary education. It has been estimated that the number of persons among the working age population 15–64 years with completed primary education was 130 million in 1970.[5] This figure may seem somewhat low, but it must be remembered that 85% of the population continues to live in rural areas and that in the past schooling has not been made available to all children. However, it should be noted that the literacy rate is likely to be considerably higher due to large-scale attempts to raise the educational level through literacy campaigns. We now make the following assumptions in order to arrive at an estimate of the number of persons having completed primary education in 1990. First, of those entering working age groups in 1971–5 60% will have completed primary schooling, in 1976–80 80%, in 1981–5 90%, and in 1986–90, 100%. The underlying assumption is, then, that primary education will become universal, even in rural areas, within the next few years.[6] The present situation is reflected in a statement by the Minister of Education at the National Education Conference: "Within three years, over 95% of all school-age children in most of the counties must attend school and be able to complete five years of schooling."[7]

The number of people having completed primary education would under these assumptions have reached about 400 million, which is approximately 60% of the working age population in 1990 against only 35% in 1970 (Table 13).

TABLE 13. *Effect of expanded primary education in China*

| Period | Yearly number of people entering working age (million) | Received primary education (%) | Total increase for period (million) | Accumulated number (million) |
|---|---|---|---|---|
| 1970 | | | | 130 |
| 1971–5 | 16 | 60 | 48 | 178 |
| 1976–80 | 17 | 80 | 68 | 246 |
| 1981–5 | 18 | 90 | 81 | 327 |
| 1986–90 | 19 | 100 | 95 | 422 |

[5] *Joint Economic Committee of Congress, People's Republic of China: An Economic Assessment,* Washington, 1972, p. 216.

[6] In an article prepared by a PRC contributor, UNESCO reports that "about 90% of school-age children are now attending school" which it is assumed relates to primary education only. Yong Hong, "The educational revolution", in *Prospects,* Vol. V, No. 4 (1975), p. 481.

In early 1976 the Chinese news agency had the following to say about primary education: "Universal five-year primary school education has been achieved in the main throughout the vast countryside of China. By the end of 1975, well over 95% of school-age children in the country had been enrolled as against 84.7% . . . in 1966, and primary school attendance had increased by 30%". (News from *Xinhua Weekly,* No. 362, London, Jan. 15, 1976).

[7] Education Minister's Report to National Education Conference (NCNA, June 10, 1978); BBC FE/5843/B11/1.

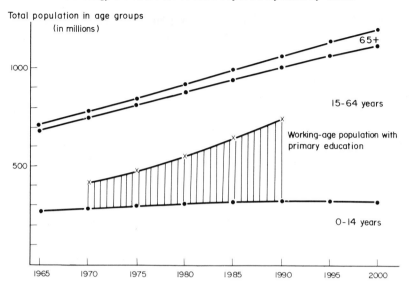

FIG. 9. *Effects of universal primary education in China.*

The figures for the working-age population having received primary education have been inserted into Fig. 9,[8] which also gives projections of China's total population in age groups over the period 1965–2000.

The fact that only 60% of the working-age population would have received primary education in 1990 – within the traditional educational system – clearly points to a continuing need for various forms of adult education. This was realized early, and there is every reason to believe that adult educational policies will be important in China's rapid economic transformation.

Expansion of primary education will have a number of effects influencing China's economic development. First, China's manpower will be increasingly better educated. Second, given the emphasis on combining theory with practice the educational expansion is likely to have a very positive effect on productivity. Third, most of the educational expansion takes place in rural areas as primary education is already universal in city areas, and, consequently, gives the countryside a strong impetus for economic development.

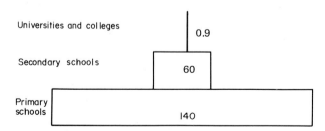

FIG. 10. *School enrolment in the People's Republic of China, 1979 (millions).*

[8] J. Sigurdson, *Comments on China's Industrial and Technological Potential in 1990,* 96 pp., National Defence Research Institute, Stockholm, 1975. (Revised, unpublished version, 1976.)

Enrolment in the primary and secondary sectors of education is given in Fig. 10. Total enrolment in universities and colleges is around 0.9 million. Consequently, only a minute fraction of a year group today goes on to higher education. There can be no doubt that China has large reserves of human talent — untapped for university and college education. Consequently, there does not appear to exist a problem in finding university entrants who are both talented and ideologically motivated.

Since the late 1960s more than 16 million educated young people have been transferred from cities and towns and more than 10 million are still working in rural areas and frontier regions.[9] As a provincial example, Guizhou reports that "In the past 10 years or so, 200,000 educated young people from various cities and townships . . . have gone to

TABLE 14. *Enrolment in primary and secondary schools in the People's Republic of China*[a, b]

|  | Primary school enrolment (million) | | Secondary school enrolment (million) |
|---|---|---|---|
| Pre-liberation peak | 24.00 | | 1.49 |
| 1949 | 24.69 | | 1.04 |
| 1958 | 86.4 | | 8.52 |
| 1966[c]   Total | (116) | 130 | (14) |
| 1973[d] | 136.80–140.73 | | 23.92–34.27 |
| 1977/8[e] | 146 | | |

[a]It has been reported that "Junior secondary schools are widespread in cities as well as in some of the rural areas. Senior secondary education has become almost universal in big cities. The number of secondary school students in 1973 was twenty-three times that before liberation, and primary school enrolment increased 5.7 fold" (Hsin Wen, "Primary and secondary education", *Prospects (UNESCO)*, Vol. 5, No. 4 (1975), p. 485).

[b]If we now turn to the statistics in *Ten Great Years*, Foreign Languages Press, Beijing, 1960, we get the following information: "In 1958 . . . the students in middle schools numbered 8,520,000 — 8.2 times the 1949 figure and 5.7 times as many as in the pre-liberation peak year. There were 86,400,000 primary school pupils in 1958, 3.5 times the number in 1949 and 3.6 times the pre-liberation peak". Surprisingly, it is also stated: "In 1958, universal primary education was put into effect in many counties; 85% of all school-age children were in school in the country as a whole."

[c]A *Red Flag* article (Dec. 1977) reports that by the time the Cultural Revolution started . . . total enrolment at schools of different categories at different levels had come to 130 million. The estimates in the table are made on the assumption that ordinary primary, and secondary schools and institutions of higher learning are included (Xinhua News Agency (Stockholm), 1977, No. 278).

[d]The variations are due to the fact that the article in *Prospects* does not clearly indicate if the base year is 1949 or a pre-liberation peak year.

[e]Most likely referring to 1977.
Rural education in a northern county (NCNA, July 1978); BBC FE/5879/B11/14.

[9] Training of middle school graduates settled in the countryside discussed at national meeting (Beijing, Jan. 24, 1978), Xinhua News Agency (Stockholm), 1978, No. 21.

the countryside". Of these, 110,000 are still working in agriculture while others have been transferred to industry, communications, finance and trade, or enrolled in cultural and educational activities.[10] There is no doubt that these educated young people represent a large, still untapped pool of resources for the development of the Chinese countryside. However, they must receive further education and training and that was one of the themes of a national conference at the beginning of 1978. The meeting decided that a spare-time education network will be set up for which the departments of education will be responsible. Other departments, like agriculture, machine-building, scientific research, and public health, should also lend a hand in training them and utilize their talents by giving technical guidance and providing special technical training. Publishing houses will also care for their special requirements in reading material.[9]

The expansion and improvement of the educational system naturally has consequences for the training and allocation of teachers. Available sources so far have only contained information about teachers in the middle and elementary schools, which comprise the 10 years of schooling before university entrance.[11] There are complaints that in the past teachers were on loan or transferred to other organizations. The inevitable result was that replacements were not available and that teaching positions were filled by non-teachers. According to a new directive the management of schools will now be under education administrative departments with the Party committees exercising centralized leadership.[12]

In order to deal with the apparent shortages, all graduates from the teachers' schools will now be assigned to educational posts. A related measure is the instruction that teachers, as well as students, should engage in industrial work, farming, and social and political activities in the time set aside for self-study during part of their vacation. This will then "ensure that teachers devote five-sixths of their time to teaching" – which is identical to the ratio given for the work performed by scientists.[12]

The same directive also discusses the commune-run schools which are likely to provide a lower quality education for some time to come. Here it is mentioned that teachers should be selected "in accordance with the actual needs for developing education and with the guidelines for maintaining the rural labour force". This is likely to mean a stronger emphasis on intellectual qualifications, and the teachers must pass an examination which is set by the education department of the county adminstration. It is also indicated that it will be difficult to change jobs without the approval of communes and schools.

### University and College Training

All countries, especially developing ones like China, have to face the challenge of training technical personnel – in appropriate numbers – to meet the various industrial and other requirements. The system of education must be coherent in order to train engineers, technicians, and other types of middle-level personnel not only in required numbers but also at the appropriate level.

Modern technology is characterized by a low toleration for error. In industrialized

---

[10] Guizhou conference on rusticated young people (Guizhou provincial service, Feb. 21, 1978); BBC FE/5749/B11/16.

[11] The total number of teachers, of all categories, is reported to be 10 million.

[12] State Council approves Education Ministry's recommendation on teachers (NCNA, Feb. 5, 1978); BBC FE/5737/B11/8.

countries much instrumentation is added to reduce the loss attributable to human error. In developing countries, where lack of experience makes human error more frequent, prevention is often achieved through close, continuous, centralized control. Consequently, it is important to train a large industrial labour force which has the necessary experience to work with modern technology. It should be noted here that very large numbers of people in China are trained in industrial technology in the numerous rural industries. This, then, enables a more decentralized approach than would otherwise have been possible.

However, advanced technology will require more high-level technicians for the following three tasks. First, to operate and maintain the new plants. Second, to evaluate, adapt designs, and duplicate the imported technology through production in Chinese machinery plants. Third, to carry out R & D of equipment and machinery to replace what is now imported or will in the near future be duplicated in China. The first two points increase the demands on the universities, while the third raises further demands on university research as well as research institutes.

It has often been argued among foreigners that China reduced her potential for industrial and technical development when the universities were closed for a number of years during the Cultural Revolution, an argument which has become more and more convincing. The stress on economic, industrial, and technological development derived from the "four modernizations" (industry, agriculture, defence, and science and technology), no doubt will greatly increase the demand for university-trained people. The long time required in the educational preparation of, for example, engineering personnel, demands that estimates of the requirements be made 5–10 years in advance. So the earlier debate on length and content of the engineering curriculum and recruiting principles clearly indicated that industrialization in the early 1980s must have been hindered. However, it can be argued that an excess number of university graduates in 1966 made it possible to close the universities temporarily, and not until recently has there been an urgent need to provide new additions to the stock of engineering personnel.

The shortage of university graduates is evident in the discussion on how to use the 160,000 graduates of the 1978 class, and the State Planning Commission made the following comments about the allocation pattern.[13] First, graduates will be assigned to key scientific research projects and large-scale construction projects. Second, priority will also be given to the selection of postgraduates and teachers of basic courses in key universities and colleges.

College and postgraduate students form one of the basic assets in the development of science and technology. The leadership in China has stated that "the quality of the new students enrolled by institutions of higher education has declined year after year, resulting in a gap in the training of scientific and technical personnel". Another criticism is that since the educational level of the students was low and their standards different, it was difficult for the teachers to teach and difficult for the students to learn. At a national conference on student enrolment held by the Ministry of Education[14] in autumn 1977, we were told that "the work of enrolling students for institutions of higher learning is

[13] College graduates assigned to construction and industry (NCNA, Aug. 15, 1978); BBC FE/5894/ B11/15.
[14] Ministry of Education's National Conference on Student Enrolment (NCNA, Oct. 20, 1977); BBC FE/5648/B11/1.

directly related to the quality of high-level specialists being trained", and that it "affects middle and primary school education".

The development of the universities in the early 1970s resulted in a serious political struggle centring on selection criteria, methods, and level of teaching, among other things. In enrolling new students it has now been made clear that the principle is to select those who are qualified while "ensuring an overall balance of moral, intellectual and physical education".[15] The national system of a unified student enrolment and distribution, practised before the Cultural Revolution, will be followed. This also means the method of voluntary application, admission of applicants by the school, and approval by the province, municipality, or region. Students are allowed to state two or three choices, noting the types of schools they wish to attend and the subjects they wish to study according to their interests and attitudes. Twenty to thirty per cent of college students will be enrolled from among senior middle school students of the current year.[16] This is clearly in contrast with the system of the post-Cultural Revolution period when students were required to work for 2 years in a factory or commune. At the time, students were also required to have party or revolutionary committee recommendation.

Some of the more important elements in the new enrolment policy are clarified in the following paragraphs.[17]

> In 1977, major reform has been carried out in the enrolment system for students of higher-learning institutes. The new methods include individual voluntary application, unified examination, preliminary selection by prefectures, municipalities, admission of applications by the school and approval by the province, municipalities and autonomous regions. All those who are qualified for enrolment, including workers, peasants, educated youth who have settled in the countryside, demobilized military personnel, cadres, and graduating high school students, are allowed to apply for the school voluntarily and to choose several schools and different academic subjects according to their own interests and abilities . . .
>
> To select students directly from among the graduating high school students for college enrolment is a directive put forward by our esteemed and beloved Premier Zhou in accordance with the guidelines of Chairman Mao's instructions. Owing to the delay and sabotage by the Gang of Four, this directive was not carried out. After the downfall of the Gang of Four, we must resolutely implement Premier Zhou's directive. By directly admitting college students from among the graduating high school students, we shall be able to enable the students to continue their study without interruption. With a most active mind and a strong memory, young people have the chance for studying fundamental theory and to become scientific research personnel. After enrolling in colleges, they should study and work in the fields as they are trained at the same time.

In admitting students, priority will be given to key colleges. Medical colleges, teacher colleges, and agricultural colleges will enrol barefoot doctors, teachers from local schools, and activists in agricultural science and technology. Here it should be observed that, aside from students from national minorities, the enrolment will also include a certain number of young people from Taiwan Province, from Hong Kong and Macao, as well as returned young overseas Chinese. After graduation they should obey the national job distribution plan decided upon by state agencies.

The examinations held in December 1977 were divided into two categories – liberal arts and natural sciences. The first includes political science, languages, mathematics, history, and geography, while the subjects of the natural sciences examination will be political science, language, mathematics, physics, and chemistry.

[15] Ministry of Education's National Conference on Student Enrolment (NCNA, Oct. 20, 1977); BBC FE/5648/B11/1.

[16] "New college enrolment system", *Beijing Review*, 1977, No. 46, p. 16.

[17] *People's Daily* on upgrading college enrolment (NCNA, Oct. 20, 1977); BBC FE/5648/B11/4.

The strong desire for and interest in university education can be realized by looking at the numbers. In the country as a whole, 5.7 million young people took college entrance examinations and 278,000 were admitted.[18] At the time of the college entrance examinations students also sat for examinations to enter secondary schools — both technical and others. At least another 10 million are likely to have sat for secondary school examinations. At the end of February 1978 the new students started to enter their colleges and universities. There can be no doubt that intellectually highly qualified students have been selected while at the same time stress was laid on the family background and a proper distribution from all over the country. At Beijing University, for example, more than 70% of the students are reported to be "children of workers, poor and lower middle peasants and revolutionary intellectuals".[19]

The geographical principle in recruiting students can be illustrated by an example from Qinghua University — China's leading technological university — where the 800 new students come from 22 of China's 30 provinces and regions including Beijing, Shanghai and Tianjin. Twenty-five were selected from Anhui, which has roughly 5% of China's population.[20] The group of key colleges to which Qinghua University belongs, has been given first priority in selecting students in order to guarantee a high standard.

The different selection criteria can also be seen from the age level of the students — again exemplified from Anhui. Here it is mentioned that Anhui Labour University and Normal University paid attention to selecting students who graduated from senior secondary schools in 1966 and 1967. Consequently, these students tend to be around 30 years old, which may be appropriate for certain types of training. On the other hand, the average age of the students selected from Anhui in 1978 by the prestigious Chinese University of Science and Technology was only 19 with 15 as the youngest and 22 as the oldest.[20]

The University of Science and Technology used to be the only high-level university of its kind — under the direct leadership of the Academy of Sciences. There are now four such universities. A second one had already been designated in Harbin, and in mid-1978 the State Council declared that the Zhejiang University should be jointly administered by both the Academy of Sciences and Zhejiang Province, particularly the former.[21] The university will continue to operate the departments of mathematics, physics, chemical engineering and optical instruments engineering, and the six other departments currently in the university. However, the university has tentative plans to open five new academic departments: calculator science, calculator engineering, material sciences, material engineering, and thermophysical engineering. The number of students enrolled this year will be double the intake of the previous year, and 2500 freshmen are expected to come from all over the country.

Zhejiang University will introduce the new academic credit system on an experimental basis. The credit system will first be applied to the new students enrolled during the second half of this year and then to other students. According to regulations, students of

[18] The educational reform and the prevention of elitism. Excerpts from interview with a leading member of the Ministry of Education (NCNA, Feb. 11, 1978); BBC FE/5739/B11/2.

[19] China's new college students begin school (Beijing, Mar. 3, 1978), Xinhua News Agency (Stockholm), 1978, No. 55.

[20] Anhui student enrolment (Anhui Provincial Service, Feb. 5, 1978); BBC FE/5739/B11/6.

[21] Zhejiang University under the Academy of Sciences (NCNA, July 24, 1978); BBC FE/5879/B11/14; and Zhejiang University's new departments (NCNA, Aug. 2, 1978); BBC FE/5883/B11/8.

institutions of higher learning adopting the academic credit system may graduate as soon as they have a satisfactory student evaluation and have the required credits. Outstanding students may be exempted from the credit requirement in subjects they have already mastered. They may chose a number of elective, new subjects, skip one or more grades, and graduate ahead of other students.

Gifted students are apparently well cared for within the University of Science and Technology.[22] The university opened a special class for 21 youngsters below the age of 16 in March 1978. New students will be admitted, and the aim is to have youngsters with the educational background of senior middle school aged 14–15. However, special arrangements to identify and foster talented young people is a feature of the new educational policy and not limited to this prestigious University of the Academy of Sciences only.

### Requirements in Engineering and Natural Sciences

We shall now use findings from India in trying to relate the stock of and estimated requirements for engineering personnel to the expansion of the Chinese economy.[23]

An Indian study[24] reveals that the *output/engineer* ratio decreases as the economy expands. The same holds for the *investment/engineer* ratio although the rate of decrease is much slower. However, the *engineer/workforce* ratio increases quite rapidly. Unfortunately, we have no estimates for the investment and the workforce in China, and we shall consequently focus on the output/engineer ratio. We have no reason to believe that the Indian economy should be structurally so different that it would invalidate our analogy.

In Table 15 we have presented the number of graduates in engineering and natural sciences year by year since 1949.[25] The accumulated number is calculated based on the yearly figures. Further, the table gives an index series for the number of graduates (100 in 1957). Table 8 also includes an industrial production index, based on a recent study of the Chinese economy (100 in 1957).[26] Further, the index series for industrial production has been projected up to 1990 — assuming growth rates ranging between 8% and 12%.

---

[22] University's special class for bright students (NCNA, July 19, 1978); BBC FE/5877/B11/14.

[23] Engineers are concerned with design, construction, and with the production and application of fundamental scientific knowledge to the problems of the physical world. Under the direction of engineers, technicians of various types supervise/perform a wide range of field operations in production and construction, testing and development, installing and running engineering plant, drafting and designing products, estimating cost and selling, and advising customers on the use of engineering or scientific equipment. Often the technician acts as liaison between engineer and skilled craftsman. It is his job to interpret the engineer's plans and designs, to determine the production and construction techniques to be used, and to choose the tools and machines best suited to the job. He is also responsible for a host of other semi-professional functions that he carries out on his own initiative and under the general supervision of a professional engineer or scientist.

[24] *Engineering Manpower – A Sectorial Study of Engineering Manpower Requirements up to 1976, Based on Output, Investment and Workforce,* Institute of Applied Manpower Research, IAMR Report No. 1/1967, New Delhi 1967. The methodology and new figures are further discussed in *Engineering Occupations in the Fifth Plan,* IAMR Report No. 1/1974.

[25] *Joint Economic Committee of Congress, People's Republic of China: An Economic Assessment,* Washington, 1972, p. 219.

[26] *China: A Reassessment of the Economy. A Compendium of Papers Submitted to the Joint Economic Committee,* Congress of the United States, Washington DC, 1975, p.

TABLE 15. *Relation between graduates in engineering and natural sciences, and industrial production in China*

| Year | Graduates (yearly average) (thousands) | Accumulated (yearly average) (thousands) | Index (1957=100) | Industrial production index (Michael Field) | | |
|------|------|------|------|------|------|------|
| 1949 | 6.3 | | | | | |
| 1950 | 6.2 | 12.5 | | | | |
| 1951 | 5.9 | 18.4 | | | | |
| 1952 | 12.4 | 30.8 | | | | |
| 1953 | 16.3 | 47.1 | | | | |
| 1954 | 16.4 | 63.5 | | | | |
| 1955 | 20.6 | 84.1 | | | | |
| 1956 | 26.0 | 110 | | | | |
| 1957 | 20.7 | 131 | 100 | 100 | | |
| 1958 | 22.1 | 153 | 117 | 145 | | |
| 1959 | (27.7) | 181 | 138 | 177 | | |
| 1960 | (57.9) | 239 | 182 | 184 | | |
| 1961 | 64.0 | 303 | 231 | 108 | | |
| 1962 | 70.0 | 373 | 285 | 114 | | |
| 1963 | 87.0 | 400 | 351 | 137 | | |
| 1964 | 86.0 | 546 | 417 | 163 | | |
| 1965 | 73.1 | 619 | 473 | 199 | | |
| 1966 | 73.1 | 692 | 528 | 231 | | |
| 1967 | 5.0 | 697 | 532 | 202 | | |
| 1968 | 10.0 | 707 | 540 | 265 | | |
| 1969 | 15.0 | 722 | 551 | 265 | | |
| 1970 | 15.0 | 736 | 562 | 313 | | |
| 1971 | 35.3 | 772 | 589 | 341 | | |
| 1972 | 35.3 | 807 | 616 | 371 | | |
| 1973 | 76.5 | 883 | 674 | 416 | | |
| 1974 | | | | 432 | | |
| | | | | Projected index (+8% +10% +12%) per year | | |
| 1975 | | | | 467 | 475 | 489 |
| 1976 | | | | 504 | 523 | 542 |
| 1977 | | | | 544 | 575 | 607 |
| 1978 | | | | 588 | 632 | 680 |
| 1979 | | | | 635 | 696 | 761 |
| 1980 | | | | 686 | 765 | 853 |
| 1981 | | | | 740 | 842 | 955 |
| 1982 | | | | 800 | 926 | 1070 |
| 1983 | | | | 864 | 1019 | 1198 |
| 1984 | | | | 933 | 1120 | 1342 |
| 1985 | | | | 1007 | 1233 | 1503 |
| 1986 | | | | 1088 | 1356 | 1683 |
| 1987 | | | | 1175 | 1491 | 1885 |
| 1988 | | | | 1269 | 1641 | 2111 |
| 1989 | | | | 1370 | 1805 | 2365 |
| 1990 | | | | 1480 | 1985 | 2648 |

The two series show that the graduate index had increased to 528 in 1966 while the industrial production index in the same year was only 231. There was in all likelihood a lack of engineering graduates in 1957, but it appears safe to assume that the continued expansion of education in engineering and natural sciences had by 1966 produced a surplus of graduates who were not required at that stage in China's industrial and economic development. The surplus of graduates may, however, be relatively small taking into consideration the fact that the output/engineer ratio is declining in a developing country like India, which should also apply to China. The assumption of an industrial growth rate of 10% per year would indicate that the number of "lost graduates" from 1967 onwards approximately corresponds to the number of "excess graduates" during the years before the Cultural Revolution. This situation is likely to have prevailed up to the early 1970s. No doubt China could have utilized the educated talent in other ways. But then it is necessary to remember that it was considered important to achieve consistency between long-term and short-term objectives and that it was necessary to establish a better basis for a development strategy in which the development of rural areas was stressed. But there can be no doubt that university education will have to expand rapidly if China decides to pursue a policy of rapid industrialization.

If we assume a yearly enrolment in colleges and universities in the region of 150,000, this would correspond to approximately 1% of the relevant age group. We now assume that university enrolment should keep pace with the assumed 10% increase of industrial production approximately equal to the rate of expansion of the modern sector of the economy. Yearly enrolment would then by 1990 have increased to approximately 690,000, which would correspond to roughly 3.5% from a base of approximately 20 million. Total university enrolment would then be about 2.8 million if we assume an average period of study of 4 years.

If we accept the projected rates of industrial growth to be between 8% and 12% as given in Table 8, it is clear that a very large number of engineering and natural sciences graduates, not counting other categories, have to be trained. Simplistic calculations indicate that China will have to train 2–4 million graduates in the period up to 1990 in order to cope with the expected demand for engineering and scientific talent within the industrial sectors, assuming the output/engineer ratio to be unchanged. In the light of the Indian experience with a declining output/engineer ratio it appears likely that the figures should be revised upwards. Further, no corrections have been made here for attrition, and the required numbers are consequently still higher. This would indicate that enrolment will have to increase considerably in existing institutions. It would most likely mean that a number of new institutions of higher learning will have to be created.

### Key Schools and Educational Opportunities

Of the almost 600 institutions of higher learning, the Ministry of Education has recently decided to list 88 as key or pilot universities for the country as a whole.[27] Sixty units already had this position before the Cultural Revolution, while the rest are recent additions to the list. Under this programme, a number of representative universities and colleges will enrol "students with a superior educational level". Such colleges will have more competent faculties, better teaching facilities, and the State will also give additional

[27] Key universities and colleges listed (NCNA, Mar. 1, 1978); BBC FE/5754/B11/13.

support for reference books and teaching equipment. The emphasis on key units in higher education is said to be justified because it is seen as "an effective means of raising the quality of education [and] to train personnel rapidly". The following universities and colleges are among the key units:

1. Beijing University
2. Fudan University in Shanghai
3. Nangai University in Tianjin
4. Nanjing University
5. Shandong University
6. Xinjiang University
7. Qinghua University in Beijing
8. Jiaodong University in Xian
9. Dongji University in Shanghai
10. University of Science and Technology
11. Beijing Aeronautical Engineering Institute
12. Daqing Petroleum Institute
13. Yunnan Forestry Institute
14. Jiangxi Communist Labour University
15. Dazhai Agricultural College

The principle of key units is not limited to the university system. At the beginning of 1978 the Ministry of Education issued a circular on a tentative programme for running key primary and middle schools throughout the country.[28] Some of the key schools will be in towns while others will be in the countryside. Colleges run by the industrial, transport, and communications enterprises will also be designated as key units. Here the stress will be on teaching basic knowledge related to the subjects in, for example, metallurgy or geology departments, and such students are also expected to do a certain amount of productive labour. The concerned educational departments expect to map out plans for their localities. In the first half of 1978 the leadership of the key units and their teaching staff will be strengthened. The key primary and middle schools will also be given financial assistance for such things as reference material and other teaching material.

It may be realistic to assume that college and university education will expand faster than industrial and economic growth. If we now assume a yearly annual enrolment increase of 15%, the yearly intake would have risen by 1990 to 1.40 million. This would then correspond to 7% of the relevant age group, and total university enrolment would amount to 5.60 million. It should also be noted that the yearly university enrolment in 1990 — assuming a yearly increase of 15% — would correspond to at least 35% of the relevant age group residing in urban areas.

On training and education Fang Yi had the following to say:

Institutions of higher learning form an important part of the scientific research front. We must speed up education and broaden the scope of its development. It is our tentative plan to make middle school education universal in the cities and junior middle school education in rural areas by the end of 1985. Great efforts will be made to develop the "July 21 colleges, the communist labour universities, spare-time education and on-the-job technical training for workers".

[28] Ministry of Education circular on primary and middle schools (NCNA, Jan. 26, 1978); BBC FE/5725/B11/6.

But due to the limitations of educational facilities, only a small number of young people will be able to enter state-run colleges. The broad masses of young people, workers, and peasants will have to rely on other types of schools to acquire and raise their scientific and cultural knowledge.[29] The July 21st worker colleges are run by industrial enterprises and mines and the May 7th colleges are training workers for rural construction. Both types of schools will draw their student body from experienced workers and peasants who are seen as an important source for the training of China's future scientific and technical personnel.[30]

The number of graduates trained annually by the ordinary state-run universities is still very limited. The July 21st worker universities, operated by local authorities, have partly made up for the shortage of graduates from colleges and universities and have filled vacancies in the technical force at plants, mines, and other enterprises (Table 16). The state council's industrial departments and some provincial, municipal, and other regional industrial departments have now formulated plans for the training of technical personnel over the next 3–8 years, until the end of the Fifth Five-year Plan (1980) and the Sixth Five-year Plan (1985) respectively. According to these plans more than half of the training tasks have been assigned to the July 21st worker universities and other schools operated by the plants themselves.[31]

The communist labour universities can be exemplified by the one existing in Jiangxi which was set up in 1958. Since then 200,000 students have graduated from the Communist Labour University, of which 170,000 have returned to their communes and production brigades as technicians or cadres. Born and bred in the countryside, they have the ideas and feelings of the people in the rural areas. The whole idea behind this type of education can be summed up in the following quotation:[32]

> Whether in enrolling students or assigning them work after graduation, the university (in Jiangxi) has never departed from the principle "From the communes and to the communes". It enrols from the communes, production brigades, state farms and farm machinery plants workers, poor and lower-middle peasants and their children who have practical experience as well as educated young people who have tempered themselves for a period of time in labour. Most return after graduation to where they have come from and a few are sent to work in agricultural scientific research institutes or other departments.

TABLE 16. *Worker Colleges (July 21st worker colleges)*

| Years | Number of colleges | Number of students enrolled |
|---|---|---|
| 1975 (June) | 1,200[a] | 90,000 |
| 1975 (December) | 5,160[b] | 250,000 |
| 1976 (March) | 6,000[b] | 460,000 |
| 1976 (July) | 15,000[a] | 780,000 |

(a)*Beijing Review*, 1976, No. 31, and *Current Scene*, 1976, No. 8, quoted from NCNA, July 21, 1976, p. 21.
(b)*Beijing Review*, 1976, No. 13, p. 23.

[29] *People's Daily* on the need for variety of educational institutions, Dec. 7, 1977; BBC FE/5688/B11/17.
[30] China holds national conference on college enrolment, Xinhua, Oct. 21, 1977.
[31] Fu Chun, An important method for training competent people – a comment on the need for plants and enterprises to run the July 21st worker universities well (NCNA, Dec. 17, 1977); BBC FE/5700/B11/5.
[32] "A new-type university", *Beijing Review*, 1977, No. 33, pp. 40–41.

The simultaneous development of rural areas and agriculture in harmony with the urban areas and industry may pose problems in education. No doubt, a balanced development is likely to involve short-term costs which, however, are more than compensated for in the long run. However, the leadership argued since the Cultural Revolution that the same educational system could serve both sectors. This is somewhat surprising in the light of the very different policies with regard to technology requirements. The refusal to accept a differentiation in the educational system can no doubt be found in the fear that new elites — a new class — would emerge in China.

This brings out a serious dilemma for the Chinese leadership. On the one hand, for obvious political and economic reasons, it is necessary to pursue a balanced economic development. On the other, the economic importance of the modern industrial sector now and in the future will require differential policies with regard to education, incentive systems, etc., which if fully implemented may tend to undermine the policies of a balanced growth. There appears to be no easy way out of this dilemma.

## Postgraduate Training

The enrolment of postgraduate students is an important measure for accelerating the training of scientific and technical talent and for realizing the "four modernizations" at an early date. These views were stated in a recent "Circular on specific methods of enrolling postgraduate students in 1977" which was issued jointly by the Chinese Academy of Sciences and the Ministry of Education. The circular calls on concerned departments to assist all units in doing a good job in registration, examination, evaluation, and selection of postgraduate students. The Academy and all schools of higher learning have begun to enrol students. The period of study will generally be 3 years. The age of those selected from university graduates must not exceed 30, while others may be admitted up to the age of 35.[33]

It has been pointed out by people within the Chinese Academy of Science that the re-institution of the postgraduate system is very important for correcting a situation where China has only a limited number of scientific research workers and those have low proficiency levels.[34] In order to advance the country's science and technology it is very important to train proficient personnel quickly.

Attempts to regularize postgraduate training have been going on for some time. Following Zhou Enlai's directives on the study of basic theories in 1972 Qinghua University in Beijing, for example set up six postgraduate classes, increased lectures on basic theories, and began improving laboratories.[35] At the time, Zhou Enlai also called for the enrolling of a part of the college students from among the new senior middle-school graduates of the year. The National Conference on College Enrolment in autumn 1977 noted that students should be encouraged to go directly from one level of education to another. This is important in training scientific researchers at a faster rate and in the study of basic theories of natural science. It is particularly important in light of the view now held in China that the political struggle in recent years "created a serious gap in the reproduction of scientific and technical personnel in all fields . . . ".[36]

[33] Enrolment of postgraduate students (NCNA, Nov. 15, 1977); BBC FE/5673/B11/12.
[34] BBC FE/5651/B11/3.
[35] Xinhua (Stockholm), 1977, No. 300.
[36] China holds national conference on college enrolment, Xinhua, Oct. 21, 1977.

PLATE 11. The photosynthesis of sand plants is being studied by the Institute of Glaciology, Frozen Soil, and Desert (Chinese Academy of Sciences) in Lanzhou. The aim is to select and popularize dune-fixing plants so that more of the desert area in the west can be brought under control.

In recent years the enrolment of research students for training has been more or less disrupted. The number of students admitted in winter 1977–8 will exceed any of the previous years, which ranged from 1000 to 3000 during the period from 1949 up to the Cultural Revolution in 1966. All universities, colleges, and institutes of the Academy of Sciences have made preparations. The Graduate School of the Chinese Science and Technology University plans to enrol 1000 postgraduate students within the next 2–3 years and then gradually increase the number.

The Chinese University of Science and Technology was founded in 1958, and the now deceased Kuo Moruo (also head of the Academy of Sciences) used to be its president.[37] During the early days, the faculty members were mainly researchers of various institutes

[37] Chinese Academy of Science and universities enrol fresh research students this winter, Xinhua (Stockholm), Oct. 22, 1977.

of the Academy who taught on a part-time basis. Now a new generation of teachers has emerged, and at present 90% of the teachers are students who graduated from Chinese universities since 1949. Since 39% of them are members of the party or the youth league, they are seen as a body of qualified teachers who are both Red and expert. Most of the nearly 8000 students who were trained and graduated at the University have reportedly made essential contributions to the development of basic sciences and modern technology. They have also participated in the successful detonation of atomic and hydrogen bombs and in successful satellite launchings. This is seen as an indication by a theoretical group of the University that the University and its teachers and graduates were supporting socialist economic development in China.[38]

Postgraduate studies at the Science and Technology University will usually last for 3 years.[39] The first year will include studies on Marxism—Leninism, natural dialectics, and other basic courses like mathematics and foreign languages. After that the students will work at various research institutes under the Chinese Academy of Sciences in the Beijing area. They will then be under the supervision of instructors who are research workers or assistant research workers. Simultaneously, they will be studying theories and engaging in practical research on specialized subjects, and before graduation will independently complete a scientific research project. The graduation project will be evaluated by the Graduate School together with the academic committee of the various units concerned. More than 200 specialities, later increased to 300, will be offered to graduate students by the Chinese Academy of Sciences.[40]

A large number of noted scientists are also working in the universities and colleges. Such institutions are seen as an important part of the scientific research work undertaken in China. Also, according to preliminary estimates, 10—20% of their teaching[41] staff are full-time scientific research workers.

The postgraduate courses at the universities will include all the basic sciences, multi-disciplinary courses like antipollution and electronic physics, the latest scientific and technological specialities, as well as philosophy and other social sciences.[42] The Zhongshan University in Guangzhou, for example, will enrol students in the following specialities: mathematics, theoretical physics, radio, optics, high polymer chemistry, organic chemistry, entymology, zoology, botany, meteorology, geology, geography, literature, history, philosophy, and ancient languages.

The new postgraduate enrolment system will accept the following categories of students:[43]

(1) graduates with political awareness and excellent scholastic standing who are physically fit and under 30 years of age (with a maximum age of 35 in exceptional cases);

(2) undergraduates with exceptionally high scholastic standing or those with scholastic standing equivalent to a university graduate;

(3) mine workers, poor and lower middle peasants, and educated youths who have

[38] Attack on the Gang's "two assessments" in education; BBC FE/5694/B11/12.
[39] BBC FE/5651/B11/3.
[40] BBC FE/5688/B11/16.
[41] "Scientific research in universities", *Beijing Review*, 1977, No. 50, pp. 30—31.
[42] Chinese Academy of Science and universities enrol fresh research students this winter, Xinhua (Stockholm 102201), Oct. 22, 1977.
[43] BBC FE/5651/B11/3.

relatively strong abilities in scientific research and who have made inventions or other creations;

(4) scientific and technological personnel in various offices, young teachers, and persons engaged in other work who may voluntarily apply for enrolment and enter only after passing an examination in politics and other difficult texts.

With a letter of introduction issued by their own units, the candidates go to the education bureau in their localities and fill out the registrations.[44] These are collected and sent to the units or institutes that the candidates want to enter. The examination is conducted in two stages. The first such examination was held at the end of February 1978 and the second 2 months later. The first examination will include a test on politics, foreign languages, basic subjects, and specialty subjects. Candidates take part in the second examination by showing the letter of notification by the Academy. The candidate also undergoes a physical examination and a political review, the latter by the unit where the candidate resides. Upon completion of this procedure the recruitment units and institutes under the Academy publish a name list of eligible candidates. Notification letters are then issued by the graduate institute of the Chinese University of Science and Technology to the eligible candidates (the first such letters were issued in May 1978).

The number of research students is still very small. In 1978 there were 57,200 candidates for approximately 9000 openings.[45] These students will be distributed among 207 colleges and universities and 144 research institutes. The Chinese interest in having large numbers of research students sent abroad for training should be seen against the still very limited capacity for advanced training at their own institutions.

With the expressed goal of increasing the number of professional research workers to 800,000, postgraduate training at universities and colleges obviously must be rapidly expanded. Qinghua University in Beijing and Zhongshan University in Guangzhou, which are both key universities, expect to have 10,000 and 5000 students respectively in 1985, which more or less corresponds to the capacity of these institutions before the Cultural Revolution. However, there are a number of new elements. Qinghua University is planning for a postgraduate/undergraduate ratio of 1 : 2, which means that it would have around 5000 postgraduate students in 1985. Thus the University would from then on provide the country annually with about 1700 students with graduate training assuming that the length of training will be 3 years. The postgraduate/undergraduate ratio at Zhongshan University is planned as 1 : 4,[46] which would give 1250 postgraduate students with about 400 graduating annually.

As an indication of the order of magnitude we assume that the number of professional research workers is at present in the region of 300,000–400,000. The burden of training the research workers must then to a considerable extent fall on the research institutes — under the academies as well as under the ministries. However, two other aspects should also be mentioned. The universities and colleges will recruit day students who obviously must be recruited from the cities where most of the universities and colleges are located. Furthermore, it is also being contemplated that universities should accept part-time graduate students. The consequence of all these measures is that the university system

---

[44] BBC FE/5688/B11/16.

[45] Enrolment of research students (NCNA, July 15, 1978); BBC FE/5875/B11/15.

[46] Information and impressions gained by the author during a visit to China in late March and early April 1978.

will be greatly expanded, far beyond the nominal enrolment capacity existing before the Cultural Revolution.

The new concern for graduate training is not limited to the natural and engineering sciences but also covers a wide spectrum of social sciences. The Chinese Academy of Social Sciences has announced that it will enrol postgraduates from the whole country. The first examinations were held in May and June 1978, and the students started their studies in September. Among the ten institutes mentioned are the Institute of Economics and the Institute of World Economy, where students will be enrolled in political economy, industrial economy, the economies of the First, Second, and Third Worlds, world statistics, etc.[47] Related to this is the resumption of *Economic Research,* a journal which was suspended for a number of years. It is being edited by the Institute of Economics under the Academy of Social Sciences, and the first issue after the resumption appeared in early 1978.[48]

Also important is the exchange of professional knowledge and experience both inside and outside the country. The revival of the Chinese Scientific and Technical Association has meant the revival of academic meetings. As an example the association sponsored an eight-day multi-disciplinary symposium in Tianjin in December 1977, which was the largest in a decade or more.[49] The symposium was attended by 500 people who were members of the societies of zoology, geography, aeronautics, metals, and forestry. Meanwhile the Society of Automation held a seminar in Beijing.

The Chinese Scientific and Technical Association was established in 1958 and now has a membership of 70 national societies dealing with natural science. All the provinces and municipalities also have their own associations with a total membership of about 1000 societies.

Another indication of the appreciation of academic excellence can be seen in the new institute committees.[50] Some institutes of the Academy of Sciences have recently set up academic committees to make fuller use of scientific and technical personnel. The Academy has already taken the formal decision to set up academic committees at the academy and institute levels. Members will be appointed on the basis of understanding the relation between socialism and science, having achieved a certain academic level, and having practical experience and scientific research specialty. The committees will primarily put forward suggestions on the orientation of scientific research, evaluate results, and assist in the training of scientific and technical personnel and review progress. Members will be appointed for 3 years by the director of the institute, approved by the institute's committee and registered with the Chinese Academy of Sciences.

The importance of studying foreign material and being able to utilize such material is stressed throughout the university system. This point has been well illustrated by Jiaodong University in Xian:

> In order to learn from foreign experience with still better results, we have stepped up collection of reference materials and gathering of scientific and technological information. We demand that all teaching seminars include in their plans of academic activities the study of the development of the related science and technology in other countries and regularly exchange ideas and experience.

[47] Social Sciences Academy to enrol postgraduate students (NCNA, Feb. 28, 1978); BBC FE/5753/B11/1.
[48] NCNA, Feb. 19, 1978; BBC FE/5748/B11/18.
[49] Xinhua News Agency, (Stockholm), 1977, No. 303.
[50] BBC FE/5673/B11/15.

Additionally, we hold short-term foreign language training classes to enable the teacher to read materials in more foreign languages. Students are also required to master one foreign language.[51]

At a meeting sponsored by the Chinese Scientific and Technical Association to prepare for the national conference it was proposed that Chinese scientists involve themselves in international changes more actively.[52] The Institute of Physics in Beijing, which in many ways seems to be at the forefront in implementing the new policies, has already drawn up plans to train researchers by sending them abroad for study and observation.[53]

[51] Institutions of higher learning must do well in scientific research by the CCP Committee of the Jiaodong University, Xian, *Red Flag,* 1977, No. 8, Selections from People's Republic of China CMP-SPRCM-77-29.

[52] NCNA, Oct. 3, 1977; BBC FE/5634/B11/6.

[53] Beijing Home Service, Sept. 29, 1977; BBC FE/5634/B11/6.

# 7. Basic and Applied Sciences

AT A conference held in October 1977 to formulate a national plan for developing basic sciences, some of the shortcomings of the past were indicated.[1] It was reported that, with the exception of some individual aspects of a few disciplines which rank among the world's best, most branches do not have enough research personnel and the level is not high enough. Furthermore, experimental equipment is backward and there are blanks and weak links.

The new outline programme for developing the basic sciences covers mathematics, physics, chemistry, astronomy, earth science, and biology. In order to implement the programme the Chinese Academy of Sciences should set up a rudimentary research network for the basic sciences within 3—5 years. This is expected to develop into a complete network covering a whole range of disciplines with modern laboratories in about 8 years — according to a national plan. It is envisaged that before the year 2000 a big proportion or the overwhelming proportion, of the various branches of the basic sciences will approach the advanced-world levels of that period, while a considerable portion will catch up with these levels and some disciplines will surpass them.

The National Science Conference identified the following eight broad areas: agriculture, energy, materials, electronic computers, lasers, space, high-energy physics, and genetic engineering — which we can assume will receive ample funds.

The Chinese Academy of Science is the national organization responsible for the promotion and development of basic sciences. In the Academy a struggle has been going on for a number of years. In an article[2] published in Shanghai in 1977 we learn that premier Zhou Enlai in 1970 instructed the Academy "to raise the level of scientific research on the basis of extensive and intensive practice, study basic theory, advance practice to the proper theoretical level, and use brilliant Mao Zedong Thought critically to inherit and develop the theory of natural science". The instructions were not fully accepted, which becomes clear from an Academy article broadcast in early 1971.

Then a revolutionary mass criticism group of the Chinese Academy of Sciences asked: "Is it necessary to integrate research in natural science with practice in production?" The members of the group answered in the affirmative and argued in the following way. They agree, they said, that it is true that most of the problems arising in production are those concerning the popularization and application of the fruits of scientific research already gained. However, they went on, many theories have been discovered in the course of popularization and application. And such theories are precisely what China needs to solve problems in industrial and agricultural production. However, they conceded that apart from the many jobs of scientific research, which are closely related to production, there are also some which are indirectly related to production today. Such is the case with jobs

---

[1] See "National plan for developing basic sciences", *Beijing Review,* 1977, No. 46, p. 3.

[2] Ko Chi, "The Gang of Four should not be permitted to strangle the theoretical study of natural science", *Wen Huibao,* July 12, 1977, broadcast by Shanghai City Service the same day; BBC FE/5569/B11/16.

of a long-term, exploratory nature. However, even here, it is pointed out that a good knowledge of production is required in order to undertake the research properly. In a summing-up statement they presented their views in the following way:

> The history of the development of natural science has fully illustrated that such development was from the very beginning determined by production. The requirements for production gave rise to natural science, and the achievements of natural science in turn guide production practice and promote production. The development of the productive forces again paves the way for further scientific development. This cycle goes on, resulting in the dialectical relationship between science and production. This is the law governing the development of science. On the other hand, if scientific research is divorced from production, it will become water without a source, or a tree without roots.[3]

Later on in the year the *People's Daily* published an article by the same writing group which still more strongly advocated the almost complete integration of scientific research with production, saying that "scientific workers should plunge into practice and go among the masses".[4] The group said that;

> Liu Shaoqi and other political swindlers like him opposed the concept that science comes from practice, peddled the sinister stuff that science comes from the brain, and vigorously asserted the academic style of research of being divorced from proletarian politics, from the worker–peasant masses and from practice. They wanted scientific research personnel to pay no attention to what went on outside their windows, and to set their minds on three things – to feel at ease in secluded dwellings, to specialize in exotic studies, and to bury themselves in their work – so as to resort to so-called sophisticated research, divorced from practice. Thus, politically, they poisoned the souls of some scientific research personnel, destroyed their knowledge and talent, and turned them into confused people without a head for politics, without affection for the workers and peasants, and without practical skill.
>
> Liu Shaoqi and other political swindlers denied that the masses of the people are the creators of history, peddled such sinister trash as that genius creates science, and advocated that scientific research units should experiment in accordance with the method used by nineteenth-century bourgeois scientists – that is, to have one or two persons experiment quietly behind closed doors. They attempted to disregard the mass movement of scientific experimentation and to exclude the broad masses of workers, peasants, and soldiers from scientific research.
>
> At some scientific research units, they let the bourgeois experts and authorities monopolise the power over scientific research. They blatantly clamoured that the basic task of a Communist Party member in the Academy of Sciences was to be good at modestly learning from the scientists; otherwise, they said, he was not worthy of being a Communist Party member. By doing this, they vainly attempted to remove the Party's leadership over scientific research.

Towards the end of the year the subject was discussed in a couple of articles in the theoretical organ of the Chinese Communist Party – the *Red Flag* – but they add nothing new to the debate.[5] It is more interesting to note that the Hunan Provincial Revolutionary Committee in 1972 held a conference on work concerning science and technology to study Chairman Mao's instructions on developing science and technology and summed up work in the province in the preceding year.[6]

---

[3] Integrate scientific research with practice in production, broadcast article (*Beijing* Home Service, Jan. 6, 1971) by the revolutionary mass criticism writing group of the Chinese Academy of Sciences; BBC FE/3586/B/5.

[4] "Science comes from practice, science comes from the masses – criticising idealist apriorism disseminated in the field of natural science by Liu Shaoqi and other political swindlers" by the mass criticism writing group of the Chinese Academy of Sciences, *People's Daily,* Aug. 14, 1971 (broadcast by Beijing Home Service, same day); BBC FE/3767/B11/1.

[5] K'o Yen, "Strive to make a success of scientific research for the revolution", *Red Flag,* 1971, No. 11, and "Persevere in the socialist orientation in conducting scientific research, investigation report on Fushun Scientific Research Institute of Coal Mining", *Red Flag,* 1971, No. 13.

[6] Hunan Conference on Science and Technology (Hunan Provincial Service, Feb. 9, 1972); BBC FE/3913/B11/1. *Hunan Daily* editorial: "Unfold scientific and technological research in a big way", Feb. 18, 1972; BBC FE/3924/B11/5.

PLATE 12. The noted mathematician Hua Legeng discusses how to apply mathematics to planning problems in order to raise efficiency (Szufang Locomotive and Rolling Stock Plant, Shandong Province).

Here we learn that the three-in-one scientific experiment groups and a scientific and technological network was to be consolidated. Throughout the reports it was stressed that in scientific experiments it was necessary to persist in the mass line and organize mass movements in a big way. The distinction between carrying out essential laboratory work in accordance with the needs of scientific research and carrying out laboratory research which is divorced from these needs was accepted. But scientific research, production, and use should be combined and (only) those revolutionary intellectuals who have integrated well with the workers, peasants, and soldiers *and* made inventions *and* created something should be praised and encouraged.

In the more recent article there is a reference to a Chinese—American physicist — presumably the Nobel prize winner Yang Chen-ning — who in 1972 volunteered the suggestion: "Do not neglect the study of basic theory. The low level of natural science theory in China has to be raised." Chairman Mao is said to have praised this opinion and Premier Zhou Enlai instructed Beijing University to run its department of natural sciences well and actively to carry out theoretical research. He also instructed science departments to grasp theoretical research.[2]

We know relatively little about the intervening period but there can be no doubt that many important research projects were dropped, e.g. work on the straton model to describe the structure of atoms, and that many theoretical study organizations continued to exist only in name or even disappeared altogether. During the period of confusion, efforts were made to discredit laboratories and set laboratory research against the practice of production in society.

It is possible to identify three phases. In 1972 the debate focused on basic research, in 1974 and again late in 1975 the debate focused on a much wider range of problems and was apparently an outcome of a struggle over the reorganization of science and technology, which is obviously related to the Fourth National People's Congress which was convened early in 1975.[7] At this time it was seriously attempted to formulate a national science and technology plan. The debate on these later occasions covered at least three major issues: (1) Who should exercise leadership of the country's research activities? (2) What was to be the proper relation and balance between production and basic research? (3) What was the relationship between philosophy and natural sciences?

The new emphasis on scientific research became evident in a number of ways including a *Red Flag* article discussing agricultural research. Here it was pointed out that it is wrong to overlook the role of special agricultural research institutes and scientists. Scientific experiments on a mass basis should be conducted. But it is also necessary to pay attention to consolidation and the building of special institutes. There is also a need to adopt proper measures to increase manpower and equipment in these institutes. The reason for this is as follows:

> Following the development of agricultural production and the mass movement for scientific experiments, specialized institutes have had to shoulder heavier tasks than before, because they have to sum up and popularize the advanced experiences of the masses, explore and discuss theory on the basis of the practice of the masses, study the problems arising in the course of current production . . . constantly conduct research into and apply advanced techniques, accumulate scientific and technical data, organize academic exchanges, and study advanced foreign techniques in a planned way.[8]

The second attack came in 1974 after the movement to criticize Lin Biao and Confucius began. The changes in 1972 were referred to as "going back to the former way of divorcing theory from practice". However, Zhou Enlai is said to have constantly pointed out to those involved in science and education that it is "essential to raise the scientific research to a higher level on the basis of extensive, deep-going practical work at the grassroots, do some basic theoretical research, and raise practice to the proper theoretical plane".[9]

A third attack came in 1975 after the Fourth National People's Congress when Zhou Enlai had put forward the task of achieving the modernization of industry, agriculture, defence, and science and technology. As a consequence he and others gave instructions on paying more attention to natural sciences and to theoretical study. They also stressed bringing specialized scientific research institutes into full play while at the same time developing scientific research among the masses.

All the attacks against the support given to basic theoretical research in natural sciences centred on the argument that "Marxist philosophy is the basic theory for all sciences", and consequently also the basic theory of natural science. The discussion at the time

---

[7] The various debates have been covered in great detail in A. Elzinga, "Red and Expert", mimeograph, 3 parts, (Institute for Theory of Science, University of Gothenburg) Gothenburg, 1977–8.
[8] Jen Wei-nung, "Strengthen research in agricultural science", *Red Flag,* 1972, No. 12; BBC FE/4176/B11/1.
[9] Ministry of Education article on Gang of Four sabotage of basic theoretical research in natural sciences, *Xinhua Weekly,* Feb. 3, 1977, London, No. 417.

reached a high degree of sophistry, which is likely to be boring to most readers. The view that Marxism is basic also in natural sciences has been refuted, and the counter-argument can be most easily understood in the following way.[10] There are two relationships. One is the relationship between natural science and philosophy. The achievements of modern natural science, exemplified by the theory of biological evolution, provide a scientific basis for the philosophy of dialectical materialism, while dialectical materialism becomes the philosophical basis of modern natural science. The other is the relationship between the applied sciences of various technologies in industry, agriculture, and medicine, on the one hand, and basic sciences like mathematics, physics, chemistry and biology, on the other. Applied sciences provide a definite technical and practical basis for basic sciences, and basic sciences in turn become the basic theories of applied science. Consequently, the basic theories of technical sciences cannot be replaced by Marxist philosophy, as was implied by the Gang of Four.

Indications of the response to the new guidelines and an awareness of the need for basic sciences is obvious in a number of articles which appeared in the second half of 1972 as is reflected in the following quotations from the September issue of *Red Flag:*

> Fundamental theoretical subjects chiefly dwell on the basic knowledge and basic theories of the laws governing the movement of matter in the world of nature and involve training in basic skills of making experiments. They play an important role in cultivating the students' ability to analyse and solve problems, prepare the essential conditions for the study of special subjects and enable the students to acquire a definite amount of knowledge in order to meet the needs of future work and continuous improvement.
> ... It is a short-sighted view which holds that fundamental theoretical subjects are useless, and one need not study such things or needs to study only a little of them.[11]

The need to relate theoretical studies to production was stressed but in somewhat modified way.

> Fundamental theoretical subjects of natural sciences which explain the basic laws of nature should be linked with production activities, reflect production needs, and mirror the demands of production practice. However, "Man's social practice is not confined to activity in production, but takes many other forms". Concepts, laws, and theories expounded in fundamental theoretical subjects of natural sciences come from the practice of production struggle and also from scientific experiments and other practical activities. More, they are closely linked with class struggle in different periods and are even a weapon for class struggle. Therefore, linking teaching of fundamental theoretical subjects with activity in production is a very important way, but not the only way, in integrating with reality.[11]

When we read today the criticism and shortcomings we have to remind ourselves that the People's Republic made great achievements in a number of areas in the period after the Cultural Revolution.[12] On April 24, 1970, China successfully launched her first earth satellite and on March 3 the following year launched a scientific satellite. Progress has no doubt been made in such areas as electronics, lasers, semiconductors, and compu-

---

[10] "A political struggle around the question of the basic theories of natural science by the mass criticism group of the Ministry of Education", *Guangming Ribao,* Jan. 16, 1977, Survey of the People's Republic of China Press, No. 6268 (CMP-SPRCP-77-4).

The article also discussed another but unrelated subject, namely the allocation of resources to projects which do not have any immediate benefits. The writers asked if "we can slight or abolish certain scientific research projects of an exploratory nature which are still at the experimental stage and do not appear to have any practical use for the time being but are of far-reaching significance?"

[11] "Make a success of teaching fundamental theoretical subjects of natural sciences" by the Educational Revolution Group of Beijing University, *Red Flag,* 1972, No. 9. Selections from China Mainland Magazines, CMP-SCMM-72-9.

[12] "Science and technology progress report", *Beijing Review,* 1973, No. 1.

ting technology to mention only some of the areas. Even if most of the work was geared to the demands arising directly out of production, this did not mean that basic theoretical research was completely neglected. Chinese scientists successfully used the method of X-ray diffraction, at very high resolution, to determine the spatial structure of a molecule of crystalline pig insulin. However, it should be remembered that the real breakthrough was the synthesis of insulin, carried out before the Cultural Revolution.

There are a number of indications that the intense debate over the integration of the professionals with the masses did not result in substantial changes of policy with regard to the high-level professionals and there were probably no fundamental changes either as far as professionals in the defence research sector were concerned. However, there can be no doubt that advanced science received little support, that many, many scientists were severely criticized within their own institutes or universities and that much work was seriously disrupted as a consequence of the political struggle.

In this vein, we refer to an inquiry[13] which tried to find out whether and to what extent *Scientia Sinica* — the major organ of the Academy of Sciences — had been influenced by the Cultural Revolution. The questions were:

(1) To what extent have the goals of the Cultural Revolution, e.g. mass participation, had consequences for the form and contents of the *Scientia Sinica?*
(2) Are there any indications that the Chinese have given up the scientific tradition in the Western sense of the word?

It must be pointed out here that this inquiry is only concerned with a single scientific journal — though it happens to be the most prestigious in China. The results based on available issues from 1973 and 1974 compared with 1965—6 clearly indicate that there is a considerable gap between mass science and advanced science. The published articles offer little indication that the elite scientists really have gone to the masses to get inspiration and ideas for solving problems. Only a couple of articles have been written by peasants and workers. There is no indication that the decentralization of the academies has resulted in advanced science being carried out at lower levels and in outlying areas of the country. As for references, most of these are still from Western sources even if there is an increase in Chinese sources. But, at the same time, there is a considerable reduction in references to Soviet sources.

The struggle over policy and the politicization of institutes could do nothing but disrupt scientific research — one of the effects being that scientists had little time left for research.

When we read about the political activities in research institutes during the period when Lin Biao was criticized we can understand that the research personnel were occasionally left with little time to do their own research. The amount of political activity may not have been much less in more recent times than in early 1973 for which the Beijing daily newspaper *Guangming Ribao* has the following to say:[14]

> Since the 2nd plenary session of the 9th Central Committee of the Party, the movement to criticize Lin Biao and rectify the style of work has developed in ever greater depth in these research units under the Chinese Academy of Sciences. More than 900 criticism meetings have been held at

---

[13] S. Dedijer and B. Billgren, "The East is read", *Nature*, Vol. 256, No. 5519 (Aug. 21, 1975); pp. 608—10.
[14] "Scientific research personnel of Chinese Academy of Sciences show seething enthusiasm for socialism during the movement to criticize Lin Biao and rectify style of work", *Guangming Ribao*, Sept. 19, 1973, Survey of China Mainland Press, No. 5469 (CMP-SCMP-73-40).

and above the level of a research office, and mass criticism meetings participated in by a few thousand people have been held thrice since the beginning of this year. The broad masses of the scientific research personnel conscientiously read the works by Marx, Lenin, and Chairman Mao, carried out revolutionary mass criticism by grasping the essence of the revisionist line pushed by swindlers of the Lin Biao type and, citing a host of facts and their personal experience, criticized their crimes in vilifying the revolutionary situation and the Party's policy towards intellectuals and opposing the road of Redness and expertness. As a result, they further enhanced their consciousness in carrying out Chairman Mao's revolutionary line.

Consequently, the circular announcing the National Science Conference to be held in spring 1978 declared that "just as we ensure the time for the workers and peasants to engage in productive labour, so the scientific research workers must be given no less than five-sixths of their work hours each week for professional work".[15]

The Physics Institute of the Academy of Sciences is one of the first institutes to announce that the principle is being implemented. The party committee says that it has taken effective measures to ensure that scientific research personnel devote at least five-sixths of their work hours each week to scientific research. The new rules are presented in the following way:

> During working hours from Monday to Friday non-professional meetings can only be held with the prior approval of the institute's party committee. When important political activities which have certain time limits must take precedence over research, the time spent on these activities must be fully made up afterwards. Except for participation in Party or CYL (Communist Youth League) activities, evening hours should generally be used as free time in which scientific research personnel can study and pursue their line of work.[16]

Another abuse, related to the inroads made on professionals' time, has been reported from Chaoyang Agricultural College and is likely to have been repeated in many other places. The college was run on the principle "Going up and going down", which originally meant that the students should return to the countryside according to the farming seasons. But the Gang of Four changed the principle to mean that the students should go to the countryside according to the "needs of the struggle". Accordingly, their tasks in returning to the countryside were changed from helping production teams to plan production and develop scientific experiments to searching out and struggling against political opponents. The college is now accused of having at the time judged the students' performance primarily on the basis of how well they struggled against political opponents. Another accusation is that the college also instituted a system of "graduation but not separation" under which the graduates were periodically summoned back to the school to report on the local situation. In various political movements the graduates were mustered and the aim was to control the students over long periods of time and gradually form a huge network. Another area of criticism — which has little to do with the balance between non-professional and professional work but emphasizes the extent of politicization — is the method of establishing close relations between the departments of the college with the corresponding bureau under the prefectural party committee. All such abuses and interferences in the professional activities of the research institutions have come to an end. This change is likely to be permanent if China proceeds on her planned road of rapid modernization which will require all the results and all the trained manpower the institutions can provide.[17]

[15] CPC Central Committee Circular on holding national science conference (Sept. 18, 19xx), *Beijing Review,* 1977, No. 40.

[16] Beijing Home Service, Dec. 12, 1977; BBC FE/5695/B11/16.

[17] NCNA reporters, A stone that was used to wildly attack the Party — Exposing and criticizing the so-called "Experience of Chao-yang Agricultural College" conducted by the Gang of Four, NCNA in Chinese, Nov. 22, 1977; BBC FE/5677/B11/8.

# 8. Mass Science

MASS science involves data gathering, study, and popularization of techniques by mobilizing large numbers of people, and is relevant in other areas than the ones discussed in the following. Meteorology and possibly geology are such areas. For example, the national network for giving forecasts and providing other meteorological data includes 16,000 commune-run weather posts which collaborate with the Central Meteorological Observatory and the weather stations.

### Agriculture

The dual character of China's science and technology system will remain, particularly in rural areas where the needs of agriculture require a system which in many ways must be different from that serving modern large-scale industry. This is likely to provide a justification for mass science for a number of years.

Today the Chinese countryside has a four-level agricultural scientific experiment network which originated in the Great Leap Forward in the late 1950s but essentially has been developed since the late 1960s. This network is composed of agricultural science institutes at the county level which, in collaboration with higher level agricultural scientific institutes, form what the Chinese call "the backbone". The other three levels are stations, teams and groups at the commune, production brigade, and production team levels respectively.

In recent years there has been a considerable expansion of the agricultural scientific experiment network which is reported to cover over half of the counties, communes, brigades, and teams in the country. This information is not very precise but we have been told that 14 million people are participating in the work of the agricultural scientific research organizations. On the basis of such information we have calculated that the manpower within this network contributed 6% of the total R & D budget in 1973 and around 10% in 1976.

The four-level (agricultural scientific) network has apparently been developed first in the big plains but has reportedly spread to most parts of the country, even if the coverage is far from complete. The network now includes activities on 2.7 million hectares of land which are used as experimental plots and fields for production of high-grade seeds. This amounts to 2.5% of the total area under cultivation or 1.7% of the sown area — which is much higher due to high rates of double and triple cropping in China.

The network is playing an important role in promoting agricultural development and also includes forestry, animal husbandry, sideline production, and fisheries, in addition to farm crops. However, one should realize that the importance of the network lies as much in solving production problems as solving problems encountered in research. In sum, the network's value lies in its ability — jointly with professional agricultural science workers — to gather data, study the breeding and cultivation techniques, and, finally, propagate the seeds over large areas. An illustration of the magnitude and complexity of the agricultural scientific network is given in Tables 17 and 18, which are based on data collected when the author visited China in 1973.

These examples of a two-leg policy with a continued emphasis on mass-based science are also supported in recent statements at the provincial level. At a mobilization meeting on science and technology in Shandong Province a high official had the following to say:[1]

Existing scientific experimentation organizations should be consolidated, strengthened and improved. Those localities and units without scientific experimentation organizations must set up such organizations within this year (1977). Efforts should be made to ensure that every factory and brigade has its own scientific experimentation organization and to ensure further that there are experimentation projects, activities, and accomplishments.

In his estimate of China's R & D expenditures in 1965, the well-known American analyst Leo Orleans states that "whereas political and social changes have permeated throughout the population, large segments of the rural population are still untouched by technological change".[2] This may by and large have been true before the Cultural Revolution, but changes since then have considerably transformed the rural sector. It is the characteristics of a basically different R & D system and the involvement of very large numbers of people rather than traditional R & D budget estimates that is the key to understanding R & D in rural China.

The personnel statistics given above are only one aspect of today's manpower with a very heavy emphasis on rural development and mobilization of large numbers of young people for technical transformation. This is evident from the fact that since the beginning of 1973 — a programme initiated earlier on a smaller scale — 2 million educated young people have every year been transferred to the countryside for a more or less permanent settlement. This corresponds to roughly 50% of school-leavers in urban areas for the period 1972–6.

But there are also problems associated with the rapid development of regional science and mass science in agriculture. The emphasis on self-reliance and the three-in-one groups of scientists, technicians, and peasants has no doubt improved the efficiency of the agricultural extension service in China. Few visitors knowledgeable about the problems of developing countries have questioned the value of breaking down the elitism and conservatism of China's scientists and bringing them into closer touch with the environment they are serving. But Dwight Perkins points out, with reference to information collected by a US Plant Studies delegation in 1974, that this new approach has cut heavily into high-level research.[3] He uses an example from the Jilin Academy where out of the 450 staff members, one-third were spending a year working on production problems at "basic points" and the other two-thirds divided their time between work in the Academy and extension work in communes that did not have "basic points".

The delegation noticed that the field research was largely straightforward variety testing. There was no sophisticated field research. There was also the question as to whether the Chinese were training sufficient numbers of highly qualified younger plant

---

[1] From a speech made by Chin Ho-chen, Secretary of the Provincial CCP Committee and Vice-chairman of the Provincial Revolutionary Committee, Jinan, Shandong Provincial Service, Sept. 24, 1977; BBC FE/5629/B11/5.

[2] L. Orleans, "Research and development in Communist China: mood, management and measurement", *An Economic Profile of Mainland China,* US Congress Joint Economic Committee, Washington, 1967.

[3] The information in the following paragraphs is mainly based on D. Perkins, "A conference on agriculture", *China Quarterly,* No. 76 (xxxxx, 1976).

*Technology and Science in the People's Republic of China*

TABLE 17. *Science and technology system in Hebei Province (1973)*[a]

|  | Hebei Province | Tangshan Region | Zunhua County |
|---|---|---|---|
| Basic statistics of the area | ≈190,000 km² 48 million (1972) 10 regions 9 cities 145 counties 3700 communes 53,000 brigades >260,000 teams | 17,600 km² 6.84 million (1972) 2 cities 12 counties 1 reclamation area ≈100 districts ≈500 communes | 1640 km² 0.58 million (1972) 1 county town 10 districts 43 communes 691 brigades 2499 teams |
| Administrative body | Science and Technology Commission | Science and Technology Group | Science and Technology Group |
|  | Planning Industry Agriculture Medicine Logistics | General    4 Agriculture  3 Industry    3 Archives & data  2 Directors   2 | General    1 Agriculture  2 Industry    1 Office     1 |
| Persons | 30 | 14 | 5 |
| S & T institutions | Province | Region | County |
|  | 150 research institutes 29 under province 39 under regions 80 under counties (Professional manpower: 7000) + 87 experimental centers 780 auxiliary stations Most counties have the same set-up of three stations plus seed farm as in Zunhua | 25 research institutes −13 under region or above −12 under counties (Farm Machinery Research Institutes) All counties have the same set-up of three stations plus seed farm as in Zunhua | 1. Farm Machinery R I A 1 Agr Technology Station A 2 Forestry Technology Station B. Veterinary Station C. Seed Station + D. Seed Farm (Professional manpower: 373, incl 75 at commune level) |
| Mass scientific enrolment | 1,000,000 (2% of population) | No enrolment figure at this level | 25,000 (4% of population) |

[a]Source: Science and technology administrative units at various levels, 1973, from *Rural Industrialization in China,* by J. Sigurdson, Harvard University Press, Cambridge, 1977, p. 86.

scientists. The members of the delegation also noted the potential problem for the Chinese scientists of being cut off or having only limited contacts with the work of agricultural scientists elsewhere in the world, and indicated that there may be little contact between plant scientists in different regions within China.

Given these shortcomings a member of the delegation suggested that:

> If you could hybridize the Indian and Chinese agricultural research and extension systems, you would have a very good system. India now has some very sophisticated research capacities and is very poor between the research organizations and the farmer. Between them and the farmer they are weak whereas between such organizations and the farmer China is strong. But China does not

TABLE 18. *Science and technology institutes in Tangshan Region, Hebei Province*[a]

1. Coal Mining Research Institute, an integral unit of the Kailuan Coal Mine Company
2. Porcelain Research Institute, an integral part of the "Porcelain Company" – physically located within one of the biggest of the approximately ten enterprises, which make up the Company
3. Chemical Industry Research Institute
4. Agricultural Research Institute, Tangshan City
5. Agro-reclaiming Research Institute, Baikezhuang
6. Fruit Tree Research Institute, Changli County
7. Aquatics Products Research Institute, Qinhuangdao City
8. Medical Research Institute, Qinhuangdao City
9. Machinery Research Institute, Tangshan City, still under preparation
10. Tangshan Region Farm Machinery and Tools Research Institute
11. Tangshan City Farm Machinery and Tools Research Institute, Tangshan City
12. Qinhuangdao Farm Machinery and Tools Research Institute, Qinhuangdao
13. Baikezhuang Agro-reclaiming Area Farm Machinery and Tools Research Institute
14. County-level Farm Machinery and Tools Research Institute in each of the twelve counties

[a]Source: Tangshan Region Science and Technology Group, 1973, from *Rural Industrialization in China,* by J. Sigurdson, Harvard University Press, Cambridge, 1977.

have anything behind it. Both of them have great gaps, and the Chinese system has a great gap in the more sophisticated work. It is probably easier to create the research part once the extension part is in place, however, than vice versa.

Most of the members of the delegation did not doubt that the Chinese could solve the problems referred to here if they wanted to. However, there was some doubt whether the Chinese political authorities were interested in moving in that direction – at least not at that time (1974). In the changes now undertaken in China, we see that many or all the problems indicated here are being observed and remedies are being implemented.

We are now told that agricultural scientific research institutions "were moved, merged or disbanded . . . at will" so that they were "deprived of the means to teach and carry on up-to-date experimentation and research".[4] The policy is now to reinstate and improve agricultural scientific research units at all levels as soon as possible, one of the consequences being that research topics should have continuity and should not be "changed at will".[4]

It has also been clearly pointed out that "scientific research is a demanding mental work and scientific researchers are also workers". On the sending of scientists to rural areas to take part in manual labour we learn that in the past working in the field in rural areas was considered as labour while working on agricultural experimental plots was not.

This criticism conforms closely with the impressions of visiting agricultural delegations from the United States and other countries. The members generally came away with a feeling that long-term research was seriously neglected while China had an excellent system for local adaptation and diffusion of agricultural technology – the latter based on the mass science concept. The latter will not be abandoned, and the *People's Daily* says in an editorial (1978):

> We must continue to consolidate and develop the four-level agricultural scientific research network with the masses' scientific research as the foundation and professional researchers and highly regard and give full play to their role as the mainstay. The results of the masses' scientific research and the experience of agricultural production models are rich resources for the professional re-

[4]*People's Daily,* 7th editorial on agricultural science (NCNA, May 7, 1978); FE/5811/B11/3.

PLATE 13. Many observers have pointed out that China may have the best agricultural science extension system of all developing countries. This is based on the enrolment of very large numbers of veteran peasants, educated youth, and a sprinkling of agronomists who have been formed into a network with local groups at four levels – the production team, the brigade, the people's commune, and the county. This network is then tied into the national system for agricultural research where in the past the long-term development tasks may have been temporarily neglected. The picture shows a scientific research group analysing rice strains in a commune in Sichuan Province.

searchers' scientific research. While scaling the heights of scientific and technological research, professional scientific researchers must also provide more theoretical and technical guidance for the masses' scientific research so that the masses can sum up their experience and improve their scientific research.[4]

Popularizing science will remain an important element in China's science and technology policy. This comes out very clearly in the CPC Central Committee Circular on Holding a National Science Conference in spring 1978. Here it is stated that "the agro-science network embracing the four levels of the county, commune, production brigade and team and the technical innovations organization should be strengthened and improved".

### Seismology

The extent of mass mobilization can be further exemplified through the work in seismology. Through the centuries China may have suffered more from earthquakes than any other country, and all have heard about the terrible disaster that struck Tangshan in Hebei Province in July 1976. No official figure for the number of casualties has been released, but the extent of damage can be understood from a report in autumn 1977, saying that 1.7 million units of earthquake-safe rooms have been rebuilt, which in English language may be an exaggeration due to the Chinese way of counting "rooms".

In 1966 it was decided that the masses should be mobilized in order to build a spare-time earthquake monitoring system to fight the earthquakes. Here we have to remember that the focus of seismological studies in China is on *prediction* in order to indicate when and where an earthquake is to take place. Compared with other earthquake-prone countries, prediction methods get more resources in China. Other countries such as the United States and Japan put more of their resources into constructing buildings which will withstand earthquakes. This may be a long-term goal in China. However, for the time being prediction is emphasized, and amateur scientists play an important role.

Seismological work in China has been studied by a number of foreign delegations, and a group of scientists from the United States have the following to say about the situation in China:

> The Chinese make general predictions about the seismicity of an area much the way Americans do. They plot the amount of energy recently released by ground motion against long-term averages. But, in contrast to American practice, the labour of making field observations is divided between professionals and "barefoot scientists" or amateurs. Professionals survey minute changes in ground elevation with laser-ranging devices which may be better than American ones. Professionals measure relative changes in the local magnetic field and sample radon levels in water. Amateurs, who may number in the tens of thousands, measure changes in the electrical resistivity of the ground, observe wells for level changes and bubbling, and note animal behaviour which could signal tremors in the earth.[5]

Today there are more than 100,000 amateur seismologists who are manning roughly 10,000 stations in addition to the 300 professionally staffed stations.[6] Most of the amateurs are teachers or technicians who carry out the seismological observations in their spare time. The stations manned by the amateur scientist vary with regard to instruments and technical quality. The most well-equipped, visited by a Swedish delegation in 1977, belonged to a school and was housed in an old Confucian temple in Kien Shui which is near Tong Hai, a place which suffered a serious earthquake in 1970. Sixty students were enrolled in the work under the supervision of the teacher. This particular unit had forty instruments of different kinds, most of them built by the school and placed in an underground tunnel. Included were a number of vertical pendulums of up to 8 metres. Observations were the basis for simple calculations, and the work related to the seismological station was seen as an application in mathematics, physics, chemistry, and geology.

The various research institutes and the large number of scientists and technicians dispersed all over the country today form a huge network. National coordination is carried out by the national seismological bureau in Beijing which has a staff of 100. The local networks are quite considerable, and Yunnan, for example, has a staff of 600 at the provincial seismological bureau and also its own network of stations.

The article by members of the Swedish delegation concludes with regard to the mass science approach in seismology that they were not told that the work of amateurs had been of critical importance in predicting earthquakes. This is more understandable when one realizes that predictions span different time scales ranging from hours to years ahead.

---

[5] "Chinese earthquakes: the Maoist approach to seismology", *Science,* Vol. 193, No. 4254 (Aug. 20, 1976), pp. 656–7.

[6] The presentation in this section is mainly based on information from *Seismologin i Kina, Kvalificerad forskning och folkrörelse.* FOA-tidningen, Vol. 15, No. 4 (Nov. 1977), pp. 3–9.

A recent *People's Daily* editorial also said that the geologists, geophysicists, and seismologists — the trained scientists — must play the major role in earthquake prediction.[7] However, the uniqueness of the Chinese approach has many dimensions. One of them is the educational effect and the fact that people engaged in prediction work may be less upset by evacuation in the case of a false alarm.

Another feature, which has wider relevance to science and technology in China, is the fact, noted by the Swedish delegation, that the evaluation of seismological measurements and almost all calculations are still being done by hand. The use of computers is very limited except for research at the larger institutes and bureaus. This is a partial explanation for the large staff of stations and laboratories compared with the situation in other countries and also underlines the national concern about the lagging development of the computer industry in China.

## Medicine

Another area of mass involvement of amateur scientists has similar characteristics to earthquake prediction work. An article in *Scientia Sinica*[8] states that esophageal cancer is one of the more common varieties of cancer in North China. During the last few years, reports from China and other countries on the epidemiology and etiology of esophageal cancer gave hints of some of the factors involved. In the same article we learn that a mass screening survey by cytological examination was the main method used for diagnosis in the areas of high incidence rates.

The large surveys recently undertaken in China would not have been possible without the active participation of the "masses", and the large retrospective survey of cancer deaths completed at the end of 1977 is such an example. By May 1977 several hundred million people had been investigated under a nationwide retrospective survey of deaths caused by malignant cancers.[9] The survey has been carried out by the National Cancer Control Office under the Ministry of Health and has, aside from the professionals, also involved a million of China's barefoot doctors — almost one-third of the total.

This is part of a nationwide programme which started in Linxian county in Henan Province, where cancer of the esophagus was prevalent. Here a team, in collaboration with research units, studied the relationship between the incidence of esophagus cancer and various factors such as consumption of mouldy corn and the extensive use of pickled vegetables in the diet. This is now one of the thirty "bases" set up throughout China which are responsible for cancer study, prevention, and treatment. The early surveys in Linxian were followed by larger surveys among the six million people of Anyang Prefecture, where the county is located, and later on among 50 million people in Henan, Hebei, and Shanxi provinces, and Beijing.

In recent years the early detection and treatment have greatly increased the rate of cure in Linxian, reports *China Reconstructs*, which also mentions that 400 barefoot doctors have been trained for the detection work. With every brigade having a barefoot

[7] Raise seismology and related technology to a new level, *People's Daily* editorial, Dec. 21, 1977 (Xinhua News Agency, Stockholm, 12 22 19).

[8] "The epidemiology of esophageal cancer in North China and preliminary results in the investigation of its etiological factors, the co-ordinating group for research on the etiology of esophageal cancer of North China," *Scientia Sinica*, Vol. XVIII, No. 1 (Jan.–Feb. 1975).

[9] Li Ping, "Cancer research in China", *China Reconstructs*, Oct. 1977, pp. 20–22.

doctor who knows how to use the apparatus, make smears, and examine the cells, early discovery has been made much easier.[9]

## Scientific and Technical Exchange

Combining the professionals and the masses is seen as one of the measures to implement the new science and technology plans. Qian Xuesen in his earlier *Red Flag* article,[10] mentions that Daqing, the well-known combine of oil wells, refineries, and petrochemical industries has a four-level scientific research network and that a four-level research network also exists in agriculture. So, he argues, it is even more necessary that the whole nation should have scientific and technological networks with level-to-level division of work and coordination. And he goes on to say that scientific and technological workers must include university, middle, and primary school-teachers and students. He also sees an expanded role for the China Scientific and Technical Association and other professional and mass organizations which should assist state scientific organizations.

Another feature of mass scientific experimental activities in China is the scientific and technical exchange stations. In 1976 such centres were reported to have been set up in 142 cities. The Shanghai Science and Technology Exchange was founded in 1970 and had 1400 members 6 years later.[11] Eighty per cent of them were industrial workers while the rest were engineers, technicians, research workers, and teachers at institutes of higher learning. The exchanges have a number of teams depending on their needs, and specialize in such fields as electronic computers. Aside from taking part in courses, the members are active among the population in the cities. They organize innovators in various branches of industry, conduct scientific experiments on a mass scale, and publicize information about new techniques.

In 1976 the system of scientific and technical exchange stations engaged altogether 1200 scientific and technical teams with a total of 54,000 people as "backbone" members. Workers made up 70% for the country as a whole with the remainder being cadres and technical personnel. This contingent of amateurs is said to take the lead in carrying out mass scientific and technical activities in the cities. Cutting across the various trades, the stations organize technical innovation activists from different factories and enterprises. Apart from exchanging advanced techniques, they also work in coordination to solve key problems in production, to train "backbone" technicians and popularize scientific knowledge by translating it into plain language and lively forms.

However, the exchange stations may have been closely associated with the technology policies of the Gang of Four and play a reduced role today. There can be no doubt that they were occasionally in conflict with the local associations for science and technology which were more professionally inclined. At the same time the emphasis of the exchange centres may have led to a relative neglect of the exchange of scientific technical information at a somewhat higher professional level. This becomes clear when reading reports that people in autumn 1977 crowded the auditorium of the Ministry of Metallurgy in Beijing to hear academic reports presented every Saturday afternoon.[12] The meetings

[10] Qian Xuesen's *Red Flag* article on science and technology; BBC FE/5563/B11/6.

[11] Workers in science and technology – the Shanghai science and technical exchange, *China Reconstructs,* Vol. 23, No. 11 (Nov. 1974). Dissemination of New Techniques, NCNA in English, Aug. 26, 1976; *BBC Summary of World Broadcasts* FE/W894/A/12.

[12] Ministry's academic report meetings, NCNA in Chinese, Dec. 13, 1977; BBC FE/5695/B11/17.

have been attended by ministers, vice-ministers, scientists, and technicians as well as political and administrative personnel. The meetings which were started in September 1977 have been sponsored by the China Metal Sciences Society and supported by the Ministry of Metallurgy. The reports during the first few months covered topics such as the iron and steel industry in the United States and Japan, advanced iron smelting, steel-refining, and steel-rolling techniques around the world. A number of the research institutes under the ministry are reported to have started holding similar report meetings. In the news report it is pointed out that the meeting hall is filled to capacity every time a meeting is held in the ministry because the lecturers are — with reference to "let a hundred schools of thought contend" — allowed to express their own views.

High-level technology, advanced scientific research, and the professionals in China are today coming back in their own right. What effect, if any, does this have on the mass science network which — as can be seen from the earlier pages — is of an impressive scale? In order to get a proper perspective we have to recognize that mass science can serve two very different functions. First, it can be used as a political instrument in order to substitute and downgrade the professionals and all intellectuals. Second, mass science can be used to complement the professional sectors where scarcity or non-competence require additional resources. Instances of the first approach will lead to the elimination of mass science or a different emphasis on it, and this problem is discussed at some length in connection with education and the role of the intellectuals in education and open-door research.

In the second approach it is possible to distinguish a number of different justifications, and the more important are the following ones. Employing large numbers of amateur scientists can be justified for collecting huge amounts of data or carrying out experiments on a large scale where high-level training is not really required. Thus, new knowledge can be created with a more efficient use of available resources. Using a large number of people in agricultural research and popularization is a good example of this approach, which may be further justified by the fact that trained manpower is not available and time is short. Involving large numbers of people in creating new knowledge — adapted to the local situation — has another advantage. It makes it easier in most cases to implement decisions where it is essential to make use of the new knowledge — introducing intercropping for example. Finally, the educational effect must not be forgotten as very large numbers of people are being trained in logical reasoning and scientifically controlled experiments and acquiring an awareness of knowledge which we take for granted in industrially advanced countries.

However, it is necessary all the time to strike a correct balance between the professionals and the people being mobilized for mass science. That a new balance is needed comes out clearly in a *People's Daily* editorial discussing earthquakes.[13] The article stresses that it is possible to obtain more information and scientific basis for analysing the forecasting of earthquakes if the masses can master a certain amount of knowledge. Pooling the efforts of the masses in observing and preventing damage caused by earthquakes is especially important because difficulties still exist in forecasting earthquakes. On the other hand, the article also says that the seismological departments, in order to push scientific research, must formulate plans for developing seismological science and

---

[13] *People's Daily* editorial on seismology, Dec. 21, 1977, "Raise seismological science and technology to a new level"; BBC FE/5701/B11/1.

technology, and also readjust and strengthen scientific research institutes. Further, it is necessary to restore the technical job titles of personnel in seismological science and technology, establish an evaluation system, and implement a system of individual rather than collective responsibility. The pattern is exemplified by the introduction of institute directors as scientific leaders.

# 9. Environmental Protection and Natural Resources

ENVIRONMENTAL pollution has generally been regarded as a consequence of economic development. The logical cure then is to expand the economy in order to pay for the required measures for reducing the pollution or else drastically reduce the rate of economic expansion or to expand the economy in a manner consistent with environmental protection. The second option is for obvious reasons not acceptable for developing countries with an urgent need to develop their national economies and gradually build up a modern industry and modern agriculture.

The present Chinese view on environment and development clearly reflects this as is apparent in the following statement:

> Economic development and environmental protection are interrelated and promote each other. The former gives rise to the environmental problem and the latter constitutes an important condition for developing the economy; economic development increases the capability to protect the environment, and environmental improvement in turn promotes economic development. This is the interdependent relationship between the two.[1]

China is industrializing at a much later stage than the now advanced countries. With a different and improved awareness of environmental needs it is likely that China can incorporate these in her development plans in order to avoid the many shortcomings in already industrialized countries.

There is also a concern for the local ecological system which has always been prominent in China, particularly with relevance to the provision of more food to meet the demands of a growing population. China's development is today characterized by decentralized industrialization, delayed urbanization, and well-designed and strong programmes to speed up rural modernization. This indicates that China's strength lies in the execution of environmental programmes. To this is added a national consensus — achieved through political information campaigns — that thrift is a virtue and that multipurpose utilization should be attempted wherever possible: in agriculture, industry, and other economic activities.

Not all pollution problems have easy solutions, as is obvious from the zoning codes which specify that certain types of industry must be located downstream and in such a way that the prevailing winds cause as little nuisance as possible in residential areas. China has inherited a substantial stock of manufacturing enterprises in the old industrial cities, and only at a later stage of economic development will such industries be brought under full environmental control.

What then are the prospects for the future? The modernization of agriculture, to increase productivity and output in order to meet the demands of the population, and to supply raw materials for industry, calls for massive application of chemical fertilizers, pesticides, and other modern inputs. The expected, continued rapid rise in the use of

---

[1] Chu, Ko-ping, "Environment and development", *Beijing Review,* No. 20 (May 14, 1976), pp. 19–20.

chemical fertilizers will sooner or later lessen China's dependence on night soil for fertilization. Furthermore, the expected industrial expansion will bring increasing risks of air and water pollution. It may be argued that China has so far been successful in her environmental concern because of her relative poverty. Domestic and international developments, and particularly the defence of the country and its natural resources, may force China to implement a rapid industrialization programme which would be more hazardous to the environment than has been the case in the past. However, none of these considerations, with the possible exception of external threats, would force China to abandon her environmental principles. There is hardly any reason why China should reverse her policy of decentralization or change her priorities and her concern for fundamental human needs.[2]

## Organizational Structure

The Stockholm Conference on the Environment (1972) is seen as the starting point for a more active phase of environmental protection in China. The State Council held its first meeting on environmental protection in 1973 and soon afterwards protection groups were set up at all relevant administrative levels. One of the first measures was to carry out a general survey of water and air pollution, and various measures have subsequently been introduced. Today all administrative levels have their own environmental protection offices in addition to the various monitoring stations which are under provincial leadership. The central agency is the Environment Protection Office (EPO) of the State Council. Its close association with investment decisions is clear from the fact that the EPO is housed in the State Capital Construction Commission in Beijing; a close relation which is also maintained at the provincial level. The main functions of the Environment Protection Agency are:

(1)  policy formulation and implementation;
(2)  formulating rules and instruction to be enforced by law;
(3)  propagate and exchange experience through lower level units;
(4)  co-ordinate environmental protection work for the whole country.

The EPO is an administrative agency, and the key centre for research is the Institute of Environmental Chemistry, previously a department of the Institute of Chemistry, under joint leadership of the EPO and the Academy of Sciences. Research is mainly focused on developing methods of analysis for various applications. In addition there exist research offices for environmental protection in a number of the Academy institutes; the Institute of Atmospheric Physics handles air pollution, the Institute of Biology and the Institute of Zoology carry out research on biological pest control. The Institute of Physics handles noise pollution, and the Institute for Water Biology deals with problems related to water pollution. In addition, the Institute of Hygiene, under the Academy of Medical Sciences, is responsible for environmental research, and several of the medical colleges are engaged in projects to find out the environmental causes of cancer and other diseases. Social factors have been mainly neglected, although collaboration with the Academy of Social Sciences, including the Institute of Philosophy and the Institute of Law, has recently

---

[2] The subject is dealt with at some length in K. W. Kapp, *Environmental Policies and Development Planning in Contemporary China and Other Essays,* Mouton Publishers, the Hague, 1974, and L. A. Orleans and R. P. Suttmeier, *Science,* Vol. 170, p. 1173 (1970).

started. Environmental protection will also be introduced in the curriculum at universities and colleges. The subject was covered at a symposium called by the Ministry of Education to discuss environmental science.[3] It was then proposed that courses and departments on environmental science will be established in a number of institutions of higher learning. The symposium also concluded that the dissemination of environmental protection ideas in primary and middle schools was important.

### Social Organization and Mobilization

China's development strategy includes two concepts which are almost lacking in most other developing countries where the approaches to economic development have been heavily influenced by models from already industrialized countries. First, China's manpower has been efficiently mobilized in almost all spheres of economic activity but in particular for rural development. Second, China has systematically utilized local resources and waste materials of all kinds in order to make the best use of the economic results. The latter approach, because of imposed frugality, has its parallel in most developing countries, although generally in a less systematic form than in China.

The mass mobilization of people has few parallels in other countries except in some of those which have adopted socialist planning. The mobilization of large numbers of people for various development projects and their active participation today has its basis in a system of well-developed local planning with a considerable degree of local control. This has created an environment which can efficiently control some of the undesired effects of economic development and provide long-term viability for modern technology.

In traditional China, the greater part of man's impact on the environment took place at the local level as a result of activities carried out by millions of peasants. Naturally, the few large environmental engineering schemes — such as flood control projects on the Yellow River — were under the central or the provincial government. For the most part, the local decisions took place in the organizational framework of the county and, at a still lower level, within the 70,000 market areas. The market area was one of the basic social and economic units in China. Each of these small units was largely self-sufficient and was linked rather weakly by trading networks to its neighbours and to higher-order economic regions.[4]

If we broaden the scope of our inquiry to include political mobilization and innovations there may be many more interesting facts to note about technological development in China.

The almost complete eradication of schistosomiasis (snail fever) which had been plaguing the population in large parts of China where people were exposed to snail-infected water is such an example. The reduction of snail fever had little to do with scientific research, nor had it much to do with new technologies. The main element of the programme to eradicate schistosomiasis is composed of two social organizational approaches — the introduction of collective units in rural areas — communes/brigades/teams — and the instrument of mobilization of large numbers of people for digging out and disposing of the mud where the snails breed. In this way it has been possible to achieve an

[3] China's institutions of higher education plan more research on environmental science (Beijing, July 11, 1978), Xinhua News Agency (Stockholm), 1978, No. 167.

[4] Much of the discussion of the local ecosystem in this section is based on the presentation in J. Whitney in *China's Developmental Experience* (M. Oksenberg, ed.), Praeger, New York, 1974.

almost complete eradication with little or no use of chemicals. The extra costs involved have also been relatively low as the work has usually been part of public works to improve irrigation and drainage in specific localities. This is an example of preventive rather than curative medicine, which is a cornerstone of China's public health system.

In the past 5 years more than 100 million people in China have taken part in the winter—spring season in construction work to improve farmland and establish irrigation and drainage facilities. These rural masses, mainly organized by brigades and communes, although co-ordinated at higher levels, have every season, on the average, been able to expand the irrigated area by more than 1.3 million hectares and have achieved flood control in another 1.3 million hectares. During the season 1975—6 still larger numbers of people are reported to have been engaged in these works and to have achieved improvements in still larger areas.

China's soil cover was extremely poor in many areas of the country when the new regime took over. In order to reduce the silt content in the rivers it was necessary to plant trees or other types of vegetation so as to prevent soil erosion or sand drifting. This has been done on a very considerable scale, particularly after the introduction of the people's communes which made it possible to mobilize large rural manpower forces to improve conditions to their own long-term benefit. The area covered by trees has almost doubled since 1949 when forests covered roughly 9% of the total land area. This would not of course have been possible if the bushes and tree saplings were collected for firewood. Consequently, the development of local coal-mines and an efficient distribution of coal at low prices has been a contributing factor.

Underlying the Chinese concepts of local planning and control of development are at least two major factors which were early recognized by the central government. First, the natural environments within each locality (region or commune) have differences which show themselves in climate, topography, vegetation, and mineral resources. Naturally these have important implications for economic and social planning. Second, there is no uniformity in the level of economic and social development and identical treatment of the localities would lead to distortions in determining the priorities.

A third, and possibly more important motivation for local planning, lies in China's development approach which stresses the improvement in the general well-being of people all over the country, access to services, culture, educational opportunities, and a high degree of participation. The meaning of development consequently includes much more than mere economic growth. So this does not mean the imposition of certain notions of development and welfare into the localities from above; the people must be involved in the decision-making process, so as to determine their own development priorities and strategies.

Here, I would like to stress that technologies have to be appropriate not only in the economic sense — in order to achieve an optimum production of goods and services; they are also required to be appropriate in the ecological sense to provide long-term viability. Furthermore, technologies, or rather the technology system, has to be appropriate in a third sense: it has also to support social and political objectives, e.g. participation.

**Useful Waste Products**

The use of various types of waste products has parallels in almost all areas of Chinese economic planning. The use of night soil as fertilizer is well-known and has until recently provided Chinese agriculture with a major share of added nutrients even though chemical fertilizer is becoming increasingly important. The newly built housing areas of the Chinese cities have flush toilets and the traditional way of collecting night soil cannot be applied. However, a number of cities, Shanghai, Beijing, and Shenyang, have constructed part of their sewer system in such a way that waste water from the housing quarters is used for irrigation and fertilization of the fields in the surrounding people's communes.

Fish breeding — in which China is a world leader — is to a considerable extent based on supplementary feeding, some of which comes from agricultural waste. Experimentation is going on to make use of fibrous vegetable matter, material which in China — as elsewhere — is available in larger quantities than the amount of grains produced. In recent years China has seen a very rapid proliferation of small units for fermentation of such agricultural waste into methane gas, as is discussed later.

The intermediate energy technologies which make considerable contributions to China's energy balance exemplify both participation and the use of residual resources.[5] About one-fourth of China's coal production comes from small pits and mines. One-third of China's hydroelectricity is generated in small plants. In the early 1970s Chinese authorities started a large and well-organized campaign to popularize the technology for using biogas. Biogas, or marsh gas as the Chinese prefer to call it, comes from fermentation in sealed pits, using animal dung, night soil, agricultural waste, or residues, etc. Aside from the value of getting a relatively cheap energy source, which can be used for cooking, lighting, and even fuel for combustion engines, this technology also improves sanitation in rural areas. The residue from the fermentation pits is finally used as fertilizer.

The popularization of the process, which is dependent on an outdoor temperature sufficiently high for fermentation, started in Sichuan. In an article on China's intermediate energy technologies, Smil calculates that the maximum potential of biogas generation in China is some 60 billion cubic metres annually, which corresponds to 50 million tons of hard coal.[5] Even if this potential cannot be fully realized, a very large proportion of the rural population lives in areas where biogas technology is viable and desirable. Smil mentions that:

> The effort is supported through national conferences (organized by the Chinese Academy of Sciences, State Planning Commission, and Ministry of Agriculture and Forestry), the training of technicians (100,000 in Sichuan alone), the manufacture of simple gas stoves and lamps, rubber on plastic pipes and pressure gauges, and through the design of differently shaped fermentation pits.

The Chinese are, no doubt, concerned with the long-term effects of using large amounts of chemicals to control agricultural pests. Large-scale experiments are carried out to find ways of replacing chemical control with biological control of pests with research being done in many places.

### Biological Pest Control

In Chinese agriculture there is still not a primary reliance on chemical methods as is the case in the United States and most of Europe. Furthermore, pest control is seen as an integral part of crop production which also exemplifies self-reliance — another characteristic of the Chinese system and practised from the national level down to the lowest collective units in the countryside. A recent visitor, Prof. H. C. Chiang,[6] describes the Chinese

[5] V. Smil, "Intermediate energy technology in China", *Bulletin of the Atomic Scientists,* Feb. 1977.
[6] H. C. Chiang, "Why China's crops have fewer pests", *Horticulture,* Jan. 1978, pp. 32—36.

system of pest control as being labour-intensive. It includes the cleaning up of plant debris, intensive monitoring to detect pest problems early, placing parasites where biological control is needed, and treating only spots of high infestation by chemical control methods. Under the conditions in China, as shown elsewhere, a substantial portion of the entire labour force can be made available for common tasks, like pest control, when necessary.

Technicians at the production level direct such activities with the support of research being carried out at the national, provincial, county, and commune levels. Chiang describes the biological control of pests by releasing disease-producing insects or parasites as an example of self-reliance. Here, county laboratories of provincial institutes maintain pure cultures of pathogens, and stocks of parasites. The communes obtain cultures from these but they do not carry out their propagation. The reason is that production in small units in the localities is easier to maintain and it is possible to make use of local material and equipment. Transportation costs will be lessened, and in all likelihood it is easier to respond to local needs both in quantity and timing and it is also possible to accept a higher level of impurity in the final product. Consequently, the monitoring of the insect populations is carried out locally by using light traps or traps with a bundle of plants and a mixture of sugar vinegar and alcohol to attract the insects. Thus, it is possible to obtain information in order to decide locally whether to treat them or not.

A quotation from Chiang's article clearly illustrates the high degree of integration of various measures which have a bearing on environmental control and the use of resources. He says that:

> ... elimination of pest insects is not an end in itself. Thus, insects caught in traps are often dumped in ponds to feed fish. Later, the muck from pond bottoms containing uneaten insects and fish metabolic wastes is hauled to the field for use as fertilizer. And, of course, the fish are then harvested. Similarly, any plant material removed from the field is put into a pile, which is then covered with paper or tarp. The insects will be driven out of the plants by solar heat but remain confined by the cover. Chickens are then herded on to these piles to feed on the insects. Finally the plant material is used to make compost. Ducks, too, are an important agent of pest control for the commune, in addition to being a production commodity. In Guangzhou, ducklings are herded into a rice paddy. In two hours they will have eaten up most of the insects in and on the water.[7]

The same observer has pointed out in a different context that the image of the Chinese ignoring basic research in entomology may be an oversimplification. In reviewing Chinese books he found the practical biological control programmes have been made possible through basic studies described in books which are comparable with publications on agricultural extension work in the United States. Chiang then highlights the priority the Chinese give to putting scientific results into operation, mentioning the publication dates. The handbook for practical use was published in 1973, while the same information in greater scientific detail was published in *Acta Entomologica Sinica,* a leading journal in this field, over a year later, in November 1974.[8] However, foreign observers have concluded that research on changes in the population of parasites — population dynamics — has at least been partly neglected with the consequence that the development of biological pest control has suffered.

[7] Chiang, op. cit., p. 34.
[8] H. C. Chiang, "Pest control in the People's Republic of China", *Science,* Vol. 192 (May 14, 1976), pp. 675–6.

**Problems**

Industrialization and rapid economic development in China cause problems similar to those experienced in other countries. A member of the Office for Environment Protection recently mentioned (1978) that the amount of industrial waste is estimated at 200 million tons annually.[9] Most of this is unutilized and either takes up space or pollutes the environment. A serious problem is the contamination of Pohai Bay which receives industrial waste liquids and discharges from the already industrialized north-east and the now rapidly industrializing provinces in North China. At the same time it was mentioned that some of the major rivers — Chang Jiang (Yangtze), the Yellow, the Huai, and the Chu — have been seriously contaminated in sections running through industrial cities because untreated industrial waste liquids are discharged into them directly or indirectly. Similar complaints are heard from Xian where the Chang River has been seriously polluted. Fifty per cent of the tributaries are reportedly filled with garbage and it is mentioned that "there are still some factories which are not paying serious attention to this issue and are still releasing poison into the river".[10]

Complaints are also heard in other places and acted upon. A spokesman for the municipal office of environmental protection in Beijing said that more than 200 written, and justified, complaints had been received last year. A number of factories were criticized for polluting the environment and the authorities were asked to take prompt action. About 100 plants under censure are in the chemical, pesticides, electroplating, iron and steel, rubber making, and textile industries.[11] The modernization of the manufacturing processes, the multipurpose utilization of waste materials and the adoption of non-toxic materials has considerably improved the situation. The factories in Beijing now release much less phenol, cyanide, chromium, mercury, and arsenic, and it is mentioned that Beijing has been relatively successful in the elimination of smog and the protection of water resources. Among still unsolved problems are the high densities of dust and sulphur dioxide in many industrial plants, far exceeding the limits set by the State. Similarly, noise is seen as a serious problem in many cities, particularly in the mining and industrial areas.

The environmental debate in industrialized countries is focused on defining and setting upper limits for toxins or polluting agents. Such limits are naturally also set in China, e.g. amounts of silica dust in mines are regulated. According to rules set already in 1956, the upper limit is 2 mg per cubic metre. Reports now mention that the dust in many mines and different workplaces is now considerably below this limit and compares favourably with the situation in other countries. This has been achieved by wet drilling, ventilation, and water spraying. Such is the situation in the State-owned enterprises, but it may still take some time until the same measures are fully introduced in the smaller collectively owned enterprises which dot the Chinese countryside.

A number of measures will be introduced in order to improve the situation.[12] A State plan, with guarantees for funds and materials, will tackle industrial pollution in stages. This will then be backed up by new laws for environmental protection, soon to be intro-

[9] Environmental Protection (NCNA, May 17, 1978); BBC FE/W982/A/3.

[10] *Xian Daily* on river pollution (Xian, Shaanxi Provincial Service, June 28, 1978); BBC FE/W988/A/4.

[11] Environmental Protection (NCNA, Apr. 15, 1978); BBC FE/W980/A/8.

[12] Interview: Environmental protection in China (Beijing, May 17, 1978), Xinhua News Agency (Stockholm), 1978, No. 120.

duced. A national environment law was promulgated in September 1979 to be followed by local regulations. More environmental research will be funded which includes the development of monitoring skills and research on basic theories of environmental science. On the question of multipurpose utilization it is realized that a single industry cannot usually by itself handle the problem; joint or co-operative efforts will consequently be supported.

## Energy

China's energy policy is multi-faceted in that the use of various energy resources are considered comprehensively. Yuan Baohua, a vice-minister of the State Planning Commission, recently had the following to say:

> The vigorous development of new techniques makes it possible to use extensive sources of energy. For example, the increase in production of coal, petroleum and natural gas, the development of hydroelectric power, the full utilization of stone coal, gangue, peat, lignite and oil shale as well as exhaust heat, the making of marsh gas, the planned development of nuclear power, research in the use of solar energy, earth's heat, wind force and tidal wave and exploration and research in other new sources of energy – all this is inseparable from new techniques.[13]

But China's energy consumption is today only roughly 0.5 ton *per capita,* in oil equivalents, which is roughly one-tenth of the consumption in highly industrialized countries. However, China has in recent years become a net exporter of oil – with Japan being the main beneficiary. With a total production approaching 100 million tons (1978), almost 10% is sold to Japan. Until recently it was often assumed that China's reserves would range from 3 to 6 thousand million tons of the world's total proven oil reserves of 100 thousand million tons. However, it has among many foreign observers been common to speculate that China's production of petroleum would increase to 400–500 million tons of crude oil per year by 1985, now discounted by Chinese planners.

Recent geological surveys – in the late 1960s – have indicated potentially rich offshore deposits in the East China Sea and the Yellow Sea. The geological structures here have theoretically been assumed to hold as much oil as the already-known deposits of the Middle East – giving China as much as 25 thousand million tons. Very little exploratory drilling has been done to prove or disprove the assertions. However, recent development of coastal oil fields in China appears to confirm the existence of geological structures rich in oil.

Consequently, China might be in a position to choose between coal and petroleum as the energy basis for her assumed rapid industrialization. Furthermore, abundant energy resources in China may support an increasing economic complementarity between China and Japan. So China might be able to meet a considerable share of Japan's demand for energy and other raw materials in exchange for capital goods needed for China's development in a number of industrial sectors. However, recently China has questioned the wisdom of shifting to oil burning in her industrial enterprises and has advocated the return to coal.

With the country's abundant water and mineral resources the main orientation is of course towards building thermo and hydroelectric power plants. However, in early 1978 in an interview with the Yugoslav news agency Tanjug, Qian Sanqiang, deputy general secretary of the Academy of Sciences, mentioned that nuclear-fuelled plants are being

[13] Vice-minister of Planning Commission on adoption of advanced technology, Xinhua News Agency (Stockholm), 1978, No. 25 (Jinan, 29, 1978).

PLATE 14. By 1985 China expects to have built ten more major oil fields and, according to Japanese forecasts (Japan External Trade Organization), produce 200 million tons annually. The picture shows construction at Zhejiang Oil Refinery.

planned, designed, and built in China.[14] China has a number of relatively small nuclear reactors and has obviously a substantial nuclear weapons programme. The emerging scientific and technical contacts clearly indicate that China is interested in gaining more knowledge about nuclear technology. However, there is no obvious reason why China, in the near future, should embark on any programme of rapid development of nuclear energy. The reasons for this are manifold. First, China's reserves of fossil fuels — petroleum and particularly coal — are sufficient to support very substantial modernization and industrialization programmes. Second, nuclear plants would by necessity be very large units and China does not yet have the required grid to distribute electricity. Third, capital

[14] BBC FE/5730/B11/12.

PLATE 15. China's continental shelf is reported to contain one of the major, still hardly explored or exploited, oil basins. Explorative work on oil and other resources is being stepped up. The picture shows scientists from the National Oceanography Bureau collecting samples of the surface layer in the East China Sea at 2000 metres.

requirements for nuclear plants would, because of capital scarcity at present, rather tend to support programmes for expanding the use of still abundant fossil fuels. Fourth, Chinese planners and politicians could argue that it is desirable to let the industrialized countries iron out all the still-too-obvious problems and pay the development costs.

In a major statement the *Beijing Review* stressed the importance of developing nuclear energy.[15] The author, Hsin Ping, in referring to international statistics, asserted that electricity generated from nuclear energy will make up 24–33% of the world's total by 2000. This indicates that "the utilization of nuclear energy can yield far greater economic results than using petroleum, coal, and natural gas. Nuclear power is opening up a broad prospect of energy sources for mankind." The Atomic Energy Institute in its plans for the future argues along the same lines:

> The application of nuclear energy resources, nuclear power, nuclear technology ... has not only had a tremendous impact on economic development, but has also had a profound effect on the nation's power and international status as well. The interrelations between atomic science and other sciences and technologies have produced many new border-line academic subjects with tremendous prospects for development.[16]

[15] "Utilization of nuclear energy and the struggle against hegemeony", Hsin Ping, *Beijing Review*, 1978, No. 15, pp. 10–12.
[16] Atomic Energy Institute calls for increased research (Beijing, Mar. 29, 1978); BBC FE/5778/C/1.

The interest in nuclear science is also reflected in the founding of a Nuclear Science and Technology Society in Shanghai. Its main function is to organize surveys and academic exchanges which will cover nuclear physics and technology, accelerators, radiation chemistry, and nuclear chemistry, among other things.[17]

However, China also advocates the need to have a more political and global view of the development of a nuclear energy capability. Hsin Ping also has a favourable impression of the Second World having made further technological progress in the use of plutonium fuel:

> The development of plutonium technology not only provides an even more effective fuel than enriched uranium, but also greatly increases the possibility of certain countries developing nuclear weapons on their own. Moreover, some countries like West Germany and France have already outstepped the United States in certain aspects of the plutonium technology. It is expected that more Second World countries and some Third World countries will be in a position to master the technology by 1985.

This is then related to the struggle for nuclear control and monopoly between the USSR and the United States. New possibilities are seen in the growing ability and need to develop a nuclear industry in the Third World, resulting in co-operation with the Second World exemplified by the West Germany—Brazil and France—Pakistan agreements.

### Mineral Resources

China is richly endowed with a wide variety of minerals including most non-ferrous metals. The resource situation, which compares favourably with that of the United States and the USSR, is due to her large land area and diversified geological structures. K. P. Wang points out that China is going through an interesting phase of geological exploration.[18] He says that mineral deposits, particularly in the south, have been examined only on the basis of traditional thinking. The plate tectonic concepts, having been recognized in recent years, with thrusts from the south moving northwards, signifies that many of the weak zones formed are conducive to mineralization and indicate future mineral discoveries.

So far the explored and exploited mineral bases of China are mostly in the north-east (Manchuria) and the west. However, with the industrialization in the past couple of decades, the north-east has declined in relative importance and the provinces in north China, with their vast coal, oil, and iron resources, have increasingly become more important. Among the non-ferrous metals, China's riches reserves are in antimony, manganese, mercury, molybden, tin, and tungsten, where reserves are either the largest or among the largest of any country. Estimated production of a few selected metals are given in Table 19.

The non-ferrous metals industry reached a high level of development in the late 1950s largely due to the use of technology imported from the USSR. In recent years the industry has shown poor performance, apparently reflecting a low priority in receiving investment funds. This has been interpreted by US analysts to be a consequence of major investment resources going to agricultural development and the petroleum, transportation, and

---

[17] Shanghai, June 29, 1978; *Xinhua News Weekly (London)*, 1978, No. 490.
[18] Most of the information in this section is taken from K. P. Wang, *Mineral Resources and Basic Industries in the People's Republic of China*, Westview Press, Boulder, 1977, and *China: The Non-ferrous Metals Industry in the 1970s — A Research Paper*, National Foreign Assessment Center, Washington DC, May 1978.

chemical fertilizer industries. Today, the Soviet technology utilized in many plants is outdated, with the result that production costs are high and yields and quality are at a

TABLE 19. *Estimated production of selected non-ferrous metals in China*

|  | 1970 | 1971 | 1972 | 1973 | 1974 | 1975 | 1976[a] |
|---|---|---|---|---|---|---|---|
|  |  |  | (thousand metric tons) |  |  |  |  |
| Aluminium | 188 | 192 | 238 | 286 | 316 | 357 | 375 |
| Antimony | 5,0 | 8,4 | 9,9 | 10,6 | 8,9 | 8,0 | 8,4 |
| Copper | 290 | 290 | 290 | 290 | 300 | 300 | 300 |
| Manganese | 1241 | 1468 | 1601 | 1772 | 1670 | 1806 | 1563 |
| Mercury[b] | 13 | 15 | 13 | 17 | 18 | 15 | 18 |
| Tin | 10 | 13 | 13 | 15 | 15 | 18 | 11 |
| Tungsten[c] | 11,5 | 16,3 | 16,6 | 18,7 | 17,1 | 14,1 | 11,9 |

[a]Preliminary.
[b]Thousand 76-pound flasks.
[c]Standard concentrates containing 60% $WO_3$.

low level. The poor performance of the sector has led to a situation where China is importing large quantities of many metals. The need to modernize the sector is recognized in the announcement of the 120 national projects of which 9 are non-ferrous complexes. In the news coverage from the National Metallurgical Conference at the beginning of 1978, the development of the steel sector is still clearly in focus.

Today the planners are concerned with low quality of iron ore deposits which require a strong emphasis on the development of benefication processes. Equally serious is the limited resources of chrome and nickel needed for special steels and the solution is seen in developing other alloy alternatives to replace the standard stainless steels based on large amounts of chrome and nickel.

# 10. Electronics

THE electronics industry has been selected as a special case to exemplify the weight which is now given to research and the support of the advanced technology sectors of the economy. A national electronics industry conference met in Beijing towards the end of 1977.[1] The day after the meeting ended, the *People's Daily* in an editorial discussed the future of the sector. Here it was categorically stated that "all branches of the national economy must be equipped with the technology of electronics before they can advance at high speed". Referring to the four modernizations, the readers were told that "the electronics industry, as an important material and technological basis for the four modernizations, should be the first to be modernized". This would then require the modernization of the science and technology required for developing components, equipment, and the manufacturing processes.

### Electronics versus Steel

However, the development of the electronics industry was seen in a very different light in the early 1970s when the sector was also under public debate. At the time the debate, which was also a critical review of suggested and/or implemented policies of the 1960s, centred on the relative importance of steel and electronics, and the outcome was that steel was the more important. This may shed some light on the reasons why electronics and other high technology sectors were — in relative terms — neglected in the intervening period.

The discussion in 1971–2 on the relative importance of the steel industry as opposed to electronic technology[2] has sometimes been interpreted as a discussion between the military and civilian sectors, with the military advocating more electronics for advanced weaponry. This is likely to be part of the explanation, but should not obscure other considerations which may be more basic to the whole development strategy of China. The electronics versus steel "controversy" can be seen as a political theme which was used to influence planners at different levels about the priorities for further development. Important areas for the use of electronic technology are data processing and automation, and it could be argued that prematurely introducing advanced electronic technology into these areas would adversely influence China's development.

Electronic technology is extensively used, for instance, for information processing, and all large companies and all planning agencies in industrialized countries make extensive use of electronic technology for planning purposes. Electronic data processing is today a

---

[1] *People's Daily,* editorial on the electronics industry: the level of the electronics industry is a hallmark of modernization (Dec. 5, 1977); BBC FE/5688/B11/2.

[2] See, for example, the *People's Daily* on relative importance of electronic industry, Beijing Radio, Aug. 12, 1971, SWB, FE/3766; Guangming Ribao, Dec. 13, 1971, "Line struggle in industry — a criticism of Liu Shaoqi's and other political swindlers' theory of 'electronics as the core' ", SCMP, No. 5045, Jan. 3, 1972.

necessity for the centralized operations of large companies. But in stressing the dispersal of industrial activities, small and medium plants were becoming increasingly important in China, and an inter-functional approach in planning was needed at lower levels. For this purpose there were positive disadvantages in using electronic data processing, since it favoured large plants (or plants under centralized control) and required highly standardized procedures. Such techniques would run directly counter to programmes for mobilizing local resources of raw materials, manpower, and savings, and would leave less leeway for local initiative.

Local industrialization which was heavily stressed for a number of years, requires a large number of machine tools for the manufacture of machinery needed for the mechanization of agriculture. Machine tools can be highly automated and integrated into groups of machines in line production by the use of electronic technology, which, in turn, reduces the need for skilled manpower. Such tools have high capacity but are not generally suitable for simple, small-scale production. Rural areas, with their developing industries, would not be ripe, either in terms of scale or level of sophistication, for the application of electronic technology.

Emphasis on electronic technology would then, among other things, mean technology for specialized, automated, high-capacity plants, which would, of course, be based in large urban centres. Machinery produced in long series could then be supplied to the countryside, but there would be few possibilities for local adaptations. As a consequence, many rural areas would be left out, either because they lacked the financial means for buying machinery or because the machinery was not suited to their conditions. The other alternative in producing machinery for the countryside has been to make relatively low-grade iron and steel available to almost all localities throughout China and to let them all gradually develop their own machinery-manufacturing capability, which is based on local skill formation and closely adapted to local land characteristics, and the financial resources available in that area. This approach in supplying the countryside with machinery requires less electronic technology, but more ingenuity in developing local steel production, from the mining of minerals to the manufacture of rolled steel.

## Concentration of Research and Production

The modern electronics industry requires a concentration not only in research but also in production. From an analysis of publishing in China it comes out that the research units involved in the compilation of books on electronics are mainly located in two large national centres, and a few smaller provincial centres, notably in Tianjin, Liaoning, Shanxi, and Hubei.[3] Comparing the geographical distribution of the research institutes to that of the production units, some interesting differences emerge.

Shanghai shows a definite concentration of electronics production units, which is not very surprising. But although Shanghai apparently is dominant in the number of production units, the number of research units in Shanghai having collaborated in publishing is almost equal to that of Beijing (Fig. 11). A natural conclusion is that Beijing gets the lion's share in research, probably on account of its status as capital. The evidence suggests that research is not so decentralized as production in the electronics sector. Therefore,

[3] E. Baark, *Dissemination Structures for Technical Information in China — An Analysis of Three Industrial Sectors: Electronics, Metallurgy, and Agricultural Machinery,* interim report (Research Policy Programme), Lund, Aug. 1978.

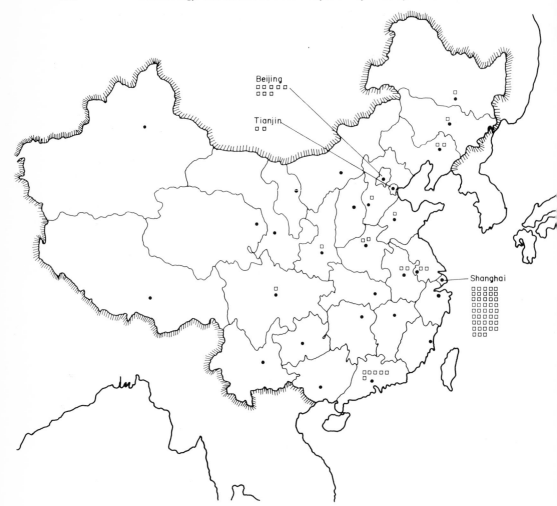

FIG. 11. *Production units involved in the compilation of books on electronics.*

although Shanghai is definitely the industrial centre with regard to production plants, research has been delegated much more evenly to the provinces.

The same study also brings out an important difference by comparing the electronics and the metallurgical sectors. While 70% of the production units in the electronics sector are located in Beijing and Shanghai, only some 25% of the production units in the metallurgical sector are located in these two centres. This is largely due to the importance of Harbin, Shenyang, and Wuhan in the metallurgical sector. In general it is apparent that research and production in the metallurgical sector is more evenly distributed compared to the electronics sector.

If one looks at the characteristics of the local engineering industry, where most of the machine tools are being used, it can be seen that the machine-tool operator controls the output of his machine. The quality of work is determined jointly by the capabilities of the machinist and of the equipment. Unless a supervisor does the job himself, there is no way to avoid the limit imposed by the expertise of individual machine operators; unless, that is, production is automated. These characteristics contrast sharply with those in the production of iron or chemical fertilizers, where most work is done by teams and the presence of a small number of knowledgeable personnel permits the operation of facilities by a work force consisting largely of unskilled labour.

Production of machinery also requires a high degree of technical skills among the work force. This explains the gradual approach being used in building up the three-level system for repair and manufacturing in rural areas, from the repair and manufacture of small simple farm machinery to the manufacture of relatively complicated machine tools. The high level of technical skills and their wide dispersion already achieved explains the capability of many counties to manufacture, for example, electric motors and diesel engines.

Heavy emphasis on the development of electronic technology would, for the time being, be likely to damage the development of more balanced socio-economic relations between cities and countryside and minimize the present growing interrelationship of the two sectors. Electronic technology introduced before extensive technical training, which has taken place all over the countryside, would be likely to increase greatly the difference in productivity between modern urban industry and rural industry, since electronic technology would then be used mainly in urban-based and relatively large plants. In addition to this, emphasis on electronic technology would be likely to reduce the availability of those planning resources needed for the transfer of technology to rural areas and the development of a local steel industry. Thus the decision to emphasize steel instead of electronics may at the time have been influenced by a desire to start extensive development of electronic technology at a later stage, when the necessary development costs can be spread more thinly on a wide economic base.

However, reports about the Chinese electronics industry showed that the planners were well aware of the potential value of electronic technology for future industrial development, although they appeared to be arguing the electronics should not be used indiscriminately.

The rapid changes in the electronics industry also have a number of thought-provoking consequences for the future. The development of micro- and mini-computer technology has today advanced very far. So China has the option of exploiting this situation to have a large share of all computer applications done on a large number of small machines instead of using a few large ones located at considerable distance from the users. So, except for certain applications, China may evade the stage of using mainly large-capacity centralized computer facilities with obvious consequences for the pattern of centralization.

## Levels in Manufacture and Prototype Development

Wang Zheng, Minister in the Fourth Ministry of Machine-Building, responsible for the electronics industry, made the following comments in an interview:[4]

> . . . within the realm of the national economy, our electronics industry is a relatively weak link, the technical level of its products is not high, its production efficiency is low, and it still cannot meet the needs of national defence and the building of the national economy. There is still a

[4] Minister interviewed on electronics industry prospects (NCNA, Nov. 16, 1977); BBC FE/5683/ B11/10.

considerable gap between the level of our electronic technology and advanced-world levels. We are not behind in the development of semiconductors, computers, and other specialized fields, but the gap between us and advanced-world levels in other areas has widened . . . .

At the production prototype level China has the expertise to do almost anything in integrated circuit technology. The expertise may be thinly spread but there can be no doubt that at the research institute level – similar to the situation in almost all research areas – the people are fully aware of the most sophisticated techniques. This is also evidenced in the news coverage from a meeting on integrated circuits held at the Metallurgical Research Institute in Shanghai in May 1978. It was reported that both Premier Hua Guofeng and Vice-premier Deng Xiaoping had attached great importance to the development of semiconductor and LSI technology. As a consequence of instructions given in the planning period before the National Science Conference, members of the Metallurgical Research Institute had completed a research project for the manufacture of two kinds of large-scale integrated circuits – apparently at the prototype stage – in approximately 6 months.[5]

Another problem is that production, in some areas, has not been able to meet the demand, with the consequence that "some units have their representatives waiting for their orders for long periods of time . . . ''. More serious may have been the poor quality of some electronics products. There is no doubt that a low ratio of up-to-standard electronic products results in a waste of raw and semi-finished materials and slows down the speed of development of the electronics industry in general. In particular, the ratio of up-to-standard integrated circuits, high-frequency and high-voltage products has been reported to be low. As an example of poor reliability and the need for repairs it is mentioned that the television sets awaiting repair account for 8% of the sets in the country. Often the maintenance and repair centres set up by the commercial departments cannot cope with the situation. Consequently, factories also have to carry out repairs, thereby slowing production.

The development of the electronics industry is by necessity closely integrated with that of the machine-building industry. Here one of the problems lies in the consolidation of the enterprises to reduce the shortcomings in production referred to earlier. The practical problems, discussed earlier, concern such matters as innovations in equipment and technology, improvements in the working environment, and enterprise management. In the *People's Daily* editorial it was pointed out that in quite a few enterprises no attention is paid to management, the division of responsibility is unclear, and no one is in charge of quality control. Some factories, the article says, have changed operational and technological procedures without authorization and brought about adverse effects on the quality of products. As a summing up the article concludes that "the most urgent task at present is to institute and improve the system of personal responsibility''. This has its parallel in the reform of institute leadership.

The components sector is likely to have been a stumbling block in the development of the Chinese electronics sector. Poor quality of components is likely to be a contributing factor in the high rate of breakdown of television sets which was referred to earlier. We can also assume that an inadequate availability of certain components, e.g. television and display tubes and LSI (large-scale integrated circuits) have hindered the growth of the electronics industry.

[5] Shanghai meeting on integrated circuits (Shanghai, June 9, 1978); BBC FE/W986/A/7.

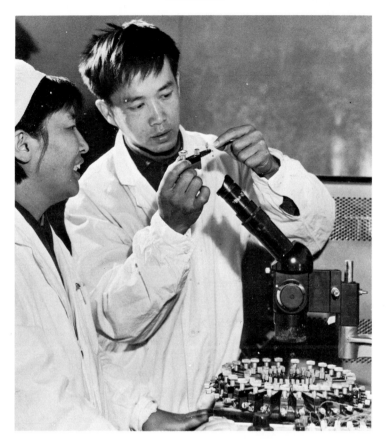

PLATE 16. Beijing and Shanghai are increasingly becoming the centres for electronics research and many of its advanced applications. The new integrated-circuits plant to be built with Japanese co-operation (Toshiba) will be constructed in Jiangsu bordering on Shanghai. The picture shows researchers at the Semiconductor Research Institute (under the Academy of Sciences) in Beijing. (Wang Hsi-ping, on the right, is a probational researcher.)

Furthermore, the fragmentation of installed capacity for a large number of products may have further aggravated the problem. The emphasis has in the past been on a wide distribution of production capacity among a large number of enterprises. The result of such a policy is that the electronics industry has been decentralized and covers a relatively large number of places. However, as a result of developments in the production techniques, the trend outside China for a number of years is towards increasing the scale of production. The likely consequences are that the electronics sector will be restructured to establish minimum viable capacity for components units assuming an increased use of foreign production technology.

**The Electronics Conference**

In connection with the 1977 Electronics Conference it has been emphasized that "compared with the advanced-world levels, China's electronics techniques are more or less backward. We must admit such backwardness. To admit is to begin to eliminate the state of backwardness, and to catch up with and surpass the advanced countries."[6] There can be no doubt that there is a strong desire in China to catch up. Why is it that electronics has become so important not only to China but to all countries that are already advanced or want to achieve modernization? The importance of this sector is presented in the following way in an article on the electronics revolution published in a *Science* special issue on electronics in 1977. Here the authors say:

> The industrial revolution, dependent on energy and materials, will be slowed and limited by the paucity of these necessary ingredients. The electronics revolution, fueled by intellectual achievements, is destined for long-continued growth as its knowledge base inevitably increases. Obviously, the current rapid rate of evolution of electronics cannot persist indefinitely, but significant change is likely to continue for a long time.[7]

The Electronics Conference in Beijing was a huge meeting with about 2500 delegates from major enterprises, scientific research units, schools of higher learning representing all provinces, municipalities, and autonomous regions of the country. Departments in charge of the industry were, of course, there. The electronics industry has already developed into quite a formidable sector with about 2800 enterprises of all sizes, spread over the country. Fifty-one research institutes more or less serving the electronics industry are attached to the Fourth Ministry of Machine-building, which is the central administrative agency in charge of the sector.[8] The development of the electronics industry is not evenly spread over the country, and the sector is concentrated around some of the major cities and in the industrially advanced areas of the country. Jiangsu Province, including Nanjing and encircling Shanghai, has, for example, 340 plants and workshops with more than 100,000 staff and workers engaged in producing electronic products. We may then conclude, assuming that the enterprises in Jiangsu are considerably larger than the national average, that the electronics industry employs more than 500,000 workers and staff.

Shanghai is seen as an important industrial base which "belongs to the people throughout China". The utilization and development of Shanghai's industry in support of industrial development in other parts of the country with manpower, material resources, and technology is discussed in a recent *Red Flag* article.[9] It is declared that the city should produce more high-quality products which are urgently needed by the State and which are technically difficult to produce. This can be done because Shanghai has the well-trained scientists and technologists, a number of scientific institutes which are materially and technically well-equipped, and long experience in organizing large, joint projects and tackling major problems.

Certain basic industries are singled out as the forms for the development of Shanghai; electronics is one of them, along with the iron and steel and petrochemical industries. Here the development and technology surrounding the major component – the electronic computer – is said to require special attention. The electronic computer technology

[6] *People's Daily*, editorial calls for rapid development of electronics industry, *Xinhua Weekly Issue (London)*, No. 460, Dec. 8, 1977.

[7] P. H. Abelson and A. L. Hammond, "The electronics revolution", *Science*, Vol. 195, No. 4283 (Mar. 18, 1977).

[8] *People's Daily*, editorial on the electronics industry: the level of the electronics industry is a hallmark of modernization (Dec. 5, 1977); BBC FE/5688/B11/2.

[9] Peng Chung, "On Development of Shanghai's industry", *Red Flag*, *BBC Summary of World Broadcasts*, FE/5691/B11/1.

should be applied extensively in all industrial, communications, capital construction, and scientific research. In further strengthening the foundation of the electronics industry, the mass production of integrated circuits and computer software has to be properly developed. If this is done, Shanghai can envisage a break-through in quantity and a tremendous leap forward not only in electronics but also in other industrial sectors as regards variety, quality, and production technology, with considerable benefits for the whole country. In this discussion the target was set for 1985, which is the end of the Sixth Five-year Plan. What is the situation like today?

In order to solve such problems and enable electronics to make its full contribution to China's modernization, the Fourth Ministry of Machine-building is mapping out a long-term programme for the development of electronic technology.[10]

The Minister points out that the development of electronics technology depends on the development of basic scientific research. Therefore, there has to be a vigorous support by the state scientific research departments. In building a comprehensive electronic scientific research system, the Minister also stresses that scientific research must be allowed to precede production and construction – an important theme of the science and technology debate which is discussed elsewhere in this book.

With reference to the international situation he also points to the need for correctly handling the relationship between study and self-invention and to study honestly and modestly all (foreign) advanced science and technology. Finally, he says that it is necessary to educate a large number of scientists and technicians for the industry. Furthermore, in addition to the professional personnel needed in the electronics sector itself, China will also need all kinds of professional personnel in related areas such as physics, chemistry, mathematics, and machine-building.

A consequence of this can be seen in the fact that electronics is also one of the subjects besides mathematics and English, which Chinese television introduced as educational courses towards the end of 1977. The courses are sponsored by the Ministry of Education and the Central Broadcasting Administration in Beijing and the broadcast programmes will also be relayed to other parts of the country.

Of the total television broadcast-time in summer 1978 on both programmes in Beijing, approximately 40% was given to education. There are certain similarities to the Open University in Great Britain, the activities of which have also been commented upon in the Chinese press. With repeat lessons, electronics covers almost 9 hours in a week of a total of approximately 48 hours. Teaching lessons are almost non-existent on Sundays when television mainly provides general information and entertainment.

## Telecommunications

China's electronic industry makes equipment for the launching and recovery of its own earth satellites which includes tracking, logging, and controlling systems. But the sector also designs and turns out telecommunications, radar, broadcasting, and television equipment, and navigational computers. All the semiconductor parts, integrated circuits, and

[10] Wang Zheng, the Minister, says that: ". . . our ideas are to grasp the vital scientific technology that is fundamental in nature and influential to the whole situation, including integrated circuit technology, solid electron technology, electronic computer technology, photo-electron technology, and space electron technology, and regard them as the major points of development . . . ". Minister interviewed on electronics industry prospects (NCNA, Nov. 16, 1977); BBC FE/5683/B11/10.

electronic elements necessary for the manufacture of such equipment can be made in China. Many kinds of data-transmission equipment have also been manufactured along with complete sets of terminal equipment. Even if China has reached a high level of manufacturing capability there is still 10% of the telecommunications equipment installed in China which has to be imported.[11]

The development of telecommunications includes priorities such as improved long distance inter-city services and extension of the national network services to more than 2000 counties, most of which already have local telephone service although of a rudimentary character. The number of telephones in the whole of China is estimated to be around 5 million (1977). Beijing, with a population of 7 million, has approximately 200,000 telephones.

Large areas in Western China are almost uninhabited, and consequently the communication problems are not easy to solve over the long distances. According to recent information, China is planning to launch two satellites in order to alleviate communications problems in the Western parts.

In spite of the existence of a national transmission plan both the power utilities and the railroads have their own communication facilities and do not use those operated by the Post and Telecommunications. Any visitor to China would have noticed that within the distance of one kilometer there may be as many as four wire pole lines parallel to the road.

At the conference on Post and Telecommunications held in July 1977 the telecommunications needs of the rural areas were strongly emphasized. The modernization of post and telecommunications will be basically completed in the country's 2000 counties by the year 1980. Circuit carrier telecommunications equipment and microwave telecommunications trunk lines will form a network. By the year 1985, postal and telecommunications departments, it is reported, will be using electronic, automatic, and mechanized equipment. Today it is only Beijing, Shanghai, Tianjin, Nanjing, and four other major cities which have adopted an automatic dialling system and still only for some of their long-distance telephone calls.

The post and telecommunications services with the outside world have also been developed. The satellite ground stations in Beijing and Shanghai have direct links with a number of countries via the international telecommunications satellites over the Indian and Pacific oceans.

Efficiency and labour productivity are also being stressed in other related areas which is obvious from the news comments on new decoding equipment.

> In September 1977 it was reported that automatic Chinese character decoding equipment had been manufactured for telegraphy. In the standard telegraphic code each of 10,000 Chinese characters is represented by a different four-digit number. The new equipment takes the four-digit code, automatically finds the corresponding Chinese character, and points it out directly.

The widespread use of this equipment in many big cities has saved a lot of manpower formerly devoted to the time-consuming task of coding and decoding by hand.

A higher level of automation of post and telecommunications is also achieved by other means. From 1 January, 1978, the Ministry of Post and Telecommunications has encouraged the authorities in Shanghai, Jiangsu, and Liaoning to implement a postal code system. After the trial period starts all mail sent to or from these three areas must bear

[11] NCNA on achievements in telecommunications; BBC FE/W951/B/1.

PLATE 17. The development of satellite communications is important for telecommunications in general and for planning in particular but has also considerable significance for the country's defence. China sent her first satellite into orbit on April 24, 1970, and many more have followed. China is tied now into the international communication satellite system. The picture shows the antenna of the first digital satellite communication station designed and built in China.

postal codes according to regulations. Under the new postal code system, a six-digit number is assigned to each of the country's tens of thousands of post offices.

The six-digit code is based on administrative districts. The first two digits represent the province or the autonomous region, the third the postal district, the fourth the county or municipality, and the last two the post office delivering the mail.

The future role of telecommunications and its application to research was made clear from a prophetic statement in Qian Xuesen's article on science and technology in *Red Flag* last summer.[12] After saying that the policy on intellectuals calls for particular attention, he goes on to discuss several professional projects and one of them is to organize the country's vast number of scientific and data units into a nationwide information and data network. Qian visualizes this as being achieved through high-density data storage and computer indexing and through the use of communications circuits, visual display terminals, and other such equipment. This would then enable scientific research personnel in any place to check or read the nation's scientific and technical documents, and promptly obtain the information and data needed from the network.

The Chinese views, or rather Qian Xuesen's personal views, on this subject are very similar to comments made in the Soviet Union in recent years:

> It is also important to organize an efficient computerized information system, which is a *sine qua non* of rational planning in science and greater efficiency of the creative scientific effort, for it helps scientists gain a knowledge of the state of Soviet and world science, of the latest scientific discoveries and technical developments, the tendencies of scientific and technical progress, and so on. It is a necessary prerequisite of the system of planning and prognostication in science and technology.
>
> Automated data-search systems are operating all over the country and new ones are being established. The long-term target is to create an automated data system for the whole country.[13]

The system envisaged is partly similar to the computer-based retrieval systems which today are linking a large number of research laboratories in western Europe, the United States, and Japan. Few of the data bases have fulfilled the early expectations, and researchers working in the front line are often better served by a personal network of contacts for the exchange of information.

## Computers

There is still a lack of usage of large computers outside the scientific circles. Foreign observers have also come to the conclusion that the Chinese computer industry has attempted to produce too many models instead of creating enough flexibility with a few general purpose models. The development of software and peripherals (computer programmes and input out devices such as readers and punchers for tapes and cards, memory storage facilities, display screens, etc.) is similar to the situation in other socialist countries and not very advanced. Programmers' work, for example, is mainly with paper tape rather than punched cards. The introduction of computers is often hindered by the very high costs for computers and peripherals which on the average are 5–10 times costlier than in the West, using the prevailing rate of exchange. Furthermore, software is poorly developed.

It can be expected that a national co-ordination programme will be worked out so that large-scale serial production of a few models can be carried out. This emphasis on large-scale production can already be noticed in other industrial sectors like farm machinery, which is one of the first sectors where reorganization measures were officially spelled out.

## Electronics and the Future

What are the reasons for the emphasis now given to the development of the electronics

---

[12] Qian Xuesen, "On science and technology", *Red Flag,* July 1977; BBC FE/5563/B11/6.

[13] *Socialism and the Scientific and Technical Revolution* (25th Congress of the Communist Party of the Soviet Union), Progress Publishers, Moscow, 1977.

PLATE 18. The electronics sector employs more than 500,000 staff and workers and is now a priority area in all aspects – research, applied sciences, and production. Computer applications are still not very common in industry. The picture shows calibration work for watches at a Shanghai factory which, by using a computer, reports an increase in productivity of 20 times for this particular operation.

industry in China? The sector cannot be considered a basic industry like energy, steel, and petrochemicals. Nonetheless, electronics constitutes an integral component – and increasingly so – in establishing a modern industrial state. The importance of the electronics sector is partly summed up in the following comments from an article discussing electronics in China:

> Communications, for example, is one indispensable aspect of any modern economy which depends entirely on electronics. Telephone, telex, telegraph, radio, television facsimile transmission, and communications satellites are electronic products without which neither civilian nor military sectors of a modern power can function.
> Electronic testing and measuring instruments are very important tools in research and development of resources and products in other industries, but they also play a vital role in the maintenance and operation of various systems and plants. Computers increase the productivity of industries and bureaucracies and are indispensable once industrialization of an economy begins its rapid growth.[14]

The Chinese view electronics more or less in the same terms. The *People's Daily* editorial referred to earlier says that all branches of the national economy must be equipped with

[14] B. O. Szuprowicz, "Electronics in China", *US China Business Review,* May–June 1976, pp. 21–43.

the technology of electronics before they can advance at high speeds. Whether in the modernization of industry, agriculture, science and technology or national defence, advanced electronic techniques must be adopted to ensure rapid growth, high quality, and high precision. Most important — at least judging from the emphasis in the article — is labour productivity. Marx's comment in discussing machines and big industry — "A well-organized system of working machines driven by a central automatic device through a transmission system is the most advanced form of production by machines" — is stressed, and seen as justification for developing the electronics industry. The " central automatic device for the well-organized system of working machines" is becoming reality in the form of control equipment with the electronic computer as the main feature. In the same discussion we also learn that China's labour productivity is still low, and the labour productivity of many departments had become even lower due to the Gang of Four's sabotage of enterprise management. This is not acceptable because, as Lenin pointed out: "Communism means the creation of a labour productivity higher than under capitalism by the workers using advanced techniques voluntarily, consciously and jointly." In the final analysis, labour productivity is the most important and primary guarantee for the victory of the new social system, the article says. And the fundamental means for raising labour productivity today is the application of the technology of electronics and other advanced sciences and technologies.

# 11. The Future

**New Technical Revolutions**

There are in various Chinese statements a number of vague references to the technical revolution when discussing China's future in science and technology. The internationally well-known scientist, Qian Xuesen, recently said that "modern science and technology is on the verge of major break-throughs", and he sees electronic computers as one of the very important areas when he says that "we are faced with new technical revolutions".[1] These quotations are from an interview with Xinhua News Agency and he says later on, with reference to Mao, that the concentration of industry and agriculture brought about by electricity at the end of the last century was a technical revolution. Further, he sees atomic energy as constituting another technical revolution. He then poses two questions: Is electronic computer technology a technical revolution? Should we actively promote this technical revolution? He does not answer the questions but, according to the Xinhua correspondent, "his view seemed to be very much in the affirmative". Both China and the Soviet Union are planned economies where the leadership base themselves on an ideology where the development of the material base is of paramount importance. It might benefit readers to have a few references to Soviet policy statements on the scientific and technical revolution, as both countries view this revolution as a key element in the future development of the material base.

In the USSR a new (second) revolution in technology is seen as paving the way for a transition from mechanization to automation — with a closer integration with science. This is one of the elements in the Soviet view of the scientific and technical revolution. Automation appears to be instrumental in the development of the advanced chemical industry, the use of micro-technology, and the development of atomic energy. There may be many similarities but also a number of distinct and important differences from this view and the views on science and technology emerging in China over the past couple of years.

The reasons for this are manifold. First, the USSR is a planned economy which has a number of superficial similarities with China. Second, the USSR stresses the significance of the scientific and technical revolutions in terms which are occasionally identical with or similar to recent statements in China. Third, some foreign observers have pointed out that the People's Republic of China will follow the roads to technological modernization already taken by industrialized countries like the USSR. Consequently, it has been decided to emphasize some of the similarities in the views on what science and technology is going to accomplish in the two countries and by what means. It is the author's hope that by pointing to similarities it will also be possible to underline the distinct differences which exist between them.

In an article on the scientific and technical revolution in the USSR, the well-known Soviet analyst Julian Cooper concludes that:

[1] Qian Xuesen on modern science and technology (Beijing, Feb. 10, 1978), Xinhua (Stockholm), 1978, No. 36.

Soviet theorists see no necessary incompatibility between the use of technique developed under capitalism and the struggle to create a socialist society. That new technical means frequently involve organizational and other social changes is not denied; what is denied is that the content and social meaning of these changes are uniquely determined by the fact that the innovations derive from a different social system.[2]

The concept of a scientific and technical revolution has been used in the USSR since the mid-1950s, and it was first referred to in the party programme in 1961. The concept has been analysed and discussed in great detail by a number of Soviet writers. Some of their views are conflicting, and the reader should refer to the full Cooper article to get a better understanding of the debate. One of the groups makes a clear distinction between the technical revolution and the production revolution, where the latter is growing out of the former following a successful social revolution. Cooper makes the following clarification:

The production revolution marks the transition to a new technological mode of production, five of which are identified: simple craft production associated with agriculture, simple craft production separate from agriculture, manufacture, machine-factory production, and comprehensively automated production characteristic of communist society.

Others differ on this and maintain that this approach is short-sighted because mechanical technology is near the limits of its potentialities and automation of such technology cannot give rise to any significant improvement in its productivity:

... the possibilities of mechanical implements of labour are nearing exhaustion, and as a result the process of transition to the application of various types of non-mechanical technology has begun. These non-mechanical forms of movement of matter (at molecular, atomic, and sub-atomic levels) can only be used technologically if control functions are transferred to technical means. This combination of non-mechanical technology with the principles of automatic control will permit the achievement of a fundamental change in the productivity of technique.

An elaboration of this point is that the scientific and technical revolution can be better understood as a revolution in the control of the natural processes involved in manufacturing processes or even as revolution in the control of processes. This is a viewpoint which comes closer to the Chinese emphasis on petrochemical industry, computer technology, and space projects.

Qian Xuesen stresses that scientists in China should move into the front lines of research, and suggests by analogy that this should include high-energy physics. He says that research in nuclear physics has given rise to a new technological revolution in the application of atomic energy.[3] So, according to him, it is very possible that high-energy physics research may give rise to yet another new technological revolution. Similarly, he sees laser and genetics engineering as promising areas where China should concentrate research and development resources of a long-range nature.

China has already announced that she expects to build a modern high-energy physics experimental base within the next 10 years. This will include a proton accelerator with a capacity of 30,000 million to 50,000 million electron volts in the first 5 years and another proton accelerator with a much bigger capacity in the remaining 5 years.[4] The number of people specialized in high-energy physics, it is said, is still fairly small.

[2] J. M. Cooper, *The Scientific and Technical Revolution in Soviet Theory in Technology and Communist Culture* (F. J. Fleron, ed.), Praeger, New York, 1977.
[3] Well-known scientist foresees technological revolution, Xinhua News Agency (Stockholm), 1978, No. 77.
[4] China to build a high-energy physics experimental base, Xinhua News Agency (Stockholm), 1978, No. 77.

However, the future will require "coordinated efforts by thousands of people under an overall plan and unified leadership. China's socialist system is well able to do this kind of collective work."[5] Accepting the concept of technical revolution and the deeper underlying changes in the manufacturing processes and in the handling of information naturally indicates an important role for the scientists who constitute a key group in promoting the technical revolution. The scientists, engineers, etc., must then be given the means required for carrying out the technical revolution. Herein lies an important element in understanding the new role of scientists.

An important element of the Soviet theory of the scientific and technical revolution is the process of the transformation of science into a direct productive force. Cooper points out that there exist in the USSR two opposing views. The first sees science as an ideal force of knowledge so that it "should be regarded as a direct productive force on its own account without the mediation of technique". The other view is that science can only become a productive force through technology or people engaged in production. Consequently, it cannot be seen as an independent element but only as knowledge materialized in the material productive forces. In summing up the scanty references made to a major article, we again quote Cooper:

> . . . all agree that science is to an ever-greater extent being transformed into a direct productive force and that the STR gives rise to profound social consequences, above all connected with the changing place of the worker in the production process. Finally, all contributors accept that the outcome of the STR depends on the nature of the social relations of the society in which it is taking place.[2]

A similar point is made by Yuan Baohua, Vice-minister of the State Planning Commission. At the national conference for exchanging experience on technical innovations in industry and communications held at Yantai, Shandong, January 15–22, 1978, he said when discussing "conservative ideas" about new technology "that ideological hindrances of all forms should be removed".[6] The exact meaning of an expression like this is still to be clarified and may not necessarily have any deep significance. But, those who have in the past been mainly concerned with ideological or political matters and have had little contact with modern science and technology, may have to do some studying in order to understand the role of science and technology.

The same minister, Yuan Baohua, indicated that implementing the technology and science programmes requires the training of those directly or indirectly concerned:[7]

> We must launch a movement to learn modern science and technology in the whole party and among the people throughout the country. In the next few years all technical personnel, cadres, and workers should receive a period of training and technical study should be made a regular practice. Cadres in leading positions should set the pace in this and take personal responsibility for popularizing new techniques.

What are the political implications, if any, of the changes in technology policy? Here we refer to Dernberger who says that

---

[5] Professor Chang Wen-yu on China's research in high-energy physics, Xinhua News Agency (Stockholm), 1978, No. 76.

[6] Vice-minister of Planning Commission on adoption of advanced technology, Xinhua News (Stockholm), 1978, No. 25 (Jinan, Jan. 29, 1978).

[7] A. S. Whiting and R. F. Dernberger, *China's Future*, Council on Foreign Relations, New York, 1977, p. 121.

PLATE 19. The development of the watch industry is one example of
the attention given to light industry and consumer goods. The picture
is from Nanning Wrist Watch Factory in Guangxi Province in South
China.

The radicals . . . using the Soviet experiences as an example, realize that the moderates' economic policies may well involve the creation and entrenchment in Chinese society of social values and
behaviour that are antagonistic to the objectives of achieving a true socialist society — so much so
that this objective is not only postponed but eventually eliminated.

## Issues of the Future

Many foreign observers have expressed the concern that the system of examination
and selection for colleges and universities will create an elite in China. In an interview
with "a leading member of the Ministry of Education", distributed by NCNA, this notion

is refuted.[8] It is admitted that, for the time being, only a limited number of students can get the opportunity to go on to university studies. But in making the selection it seems natural to use examinations, and priority is given to worker—peasant candidates when their qualifications are more or less similar. Certain specialized institutions like agricultural, medical, and teacher colleges "pay attention to admitting agrotechnical enthusiasts, barefoot doctors, and local school teachers". In addition, some colleges undertake the training of people for rural communes with students going back to their communes after graduation.

Another important fact is that wages are low for the "millions of qualified college students" who have been trained after 1949. Their wages are almost the same or only slightly higher than their worker counterparts, with at present less than 100 yuan for the earliest graduates of the period and around 40 yuan for the more recent graduates.[8] The average wage for industrial workers is in the region of 60 yuan with a ratio of 3 to 1 between highest and lowest.

The question of elitism and the possibility that elite groups in the Chinese society may develop into privileged groups is naturally related to the question of urban—rural relations, the rates of development in industry and agriculture, and the terms of trade between these two major sectors of the economy. However, more important may be the rapid development of the quaternary sector — in various ways handling and processing information — and the attitudes of those who make up that sector. Within the sector is found a greater proportion of decision-making activity, administrative service, and research. This fourth sector can be seen in parallel to the other three major sectors — raw materials production, manufacturing, and services. The sector plays an important role in all advanced economies because success in most economic fields is dependent on administrative expertise particularly in the development of industrial goods. Even the production of public goods and services increasingly requires specialist administrative guidance and co-ordination.

The tasks carried out within the quaternary sector can also be categorized as higher level administration and would include formulation of ideas, exchange of ideas, exchange and processing of information, planning, management, and co-ordination.[9] It can be seen that those engaged in such activities make up a large proportion of the "intellectual" group. Many of them may only vaguely relate to workers and peasants except through family background. With an increasing emphasis on professionalism and specialization and a subsequent reduction in requirements for manual labour, many of those who make up the quaternary sector may tend to look upon themselves as superior. Compounding the problem is the fact that specialization usually leads to centralization, although much of the information needed for decision-making can readily be transferred using modern telecommunications. However, the qualitatively important parts will still require individuals to travel and meet, the consequence of which is that very high levels of people's interaction is found in a few dominant places within a country. This clearly points to centralization, a consequence of the need for key individuals to minimize the time needed for travel and external contacts.

[8] The educational reforms and prevention of elitism, NCNA, Feb. 11, 1978; BBC FE/5739/B11/2.
[9] I am indebted to Tommy Carlstein for sharing a preprint copy of "The study of activities in the quaternary sector", by Lars-Olof Olander and Tommy Carlstein, to appear in T. Carlstein, D. Parkes, and N. Thrift, eds., *Human Activity and Time Geography, Timing Space and Spacing Time*, Vol. 2, Edward Arnold, London.

The development of industrial enterprises in rural areas — in particular the collectively owned ones which have proliferated within communes and brigades in recent years — poses a number of problems with regard to equality.[10] The wages paid to the workers in such industries are with few exceptions higher than what the agricultural labourers receive in the area surrounding the industrial units. In a news item publicizing the experience of a brigade in Guangdong it was stressed that "payment from brigade-run enterprises for each person transferred from a production team should be given to the individual's production team". It is then incorporated with the team's work points to be distributed among all the commune members, including those who work in the brigade-run industries. Whether this policy will be upheld nationwide or not will not be discussed here. However, there is a compelling logic in view of the low level of mechanization and the rapid development of brigade- and commune-run industries. The most important asset of the production team is its labour force, and the indicated policy assures that the remuneration for this asset — if used outside the production team — still benefits the members of the team and not the individuals who are transferred.

However, there is also another problem with regard to the small-scale industries in rural areas. The labour productivity is much lower in the collectively owned enterprises than in the state-owned rural industries, which in turn is considerably lower than in the large-scale state-owned enterprises, mainly located in the urban areas. In terms of average production value the ratio is $2:5:10$ in favour of the large state-owned enterprises — according to estimates made by the author. The lower labour productivity is a reflection of quality of management, the amount of capital, and the technology utilized. If capital and technological resources are allocated on a priority basis to the large state-owned enterprises the consequence might be that the productivity of the smaller, rural enterprises would remain low — in relative terms. This may not have been a clear objective in the present strengthening of management and central initiative. However, in discussing technical innovations at a national conference, Yuan Baohua, Vice-minister of the State Planning Commission, said:

> ...in the system of industrial management, there must be planned reform to effect specialized production. It is not right for us to build large or small factories that engage in all kinds of specialities. New technologies are recommended only for large-scale, specialized mass production.[11]

Similar views, which might be detrimental to local industrial development, were also voiced when discussing policies for the mechanization of agriculture at the beginning of 1978, referred to earlier.

A related question is the need to standardize industrial products and produce them on a large scale. Farm machinery is manufactured in a large number of plants, and the call to stop "the disorderly production of some farm machinery" no doubt is likely to mean that part of the rural industrial structure will be reorganized. This relates particularly to the engineering sector of local industries and it is now declared that "The problem where parts for farm machines of identical numbers are not interchangeable must be solved". This is, of course, a clear indication that the sector will receive much more attention from the central authorities and co-ordination measures will be implemented in order to

[10] Production team as basic unit in three-level ownership, NCNA, Feb. 13, 1978; BBC FE/5743/B11/12.
[11] Vice-minister Yuan's speech at Technical Innovations Conference, NCNA, Feb. 3, 1978; BBC FE/5739/B11/9.

PLATE 20. Local industry is an essential and dynamic element in China's economic policy. The picture, from <u>Yipin</u> Chemical Works in Sichuan, where workers are filling hydrochloric acid, indicates that many operations are still labour-intensive and cumbersome.

organize farm machinery for mass production. The resulting changes will, it is hoped, make it possible to "raise the quality of products, cut production costs, and make operation and maintenance easy in the countryside".[1 2]

Finally, we turn to a couple of issues which have been hotly debated during the past couple of years, one of which has been the role of basic, theoretical, and long-term research. As long as China had an untapped potential of knowledge and resources for economic development, this may not have been an important question. But foreign observers and the Chinese themselves clearly see that this is no longer the case, though the gap between China and advanced countries may be widening in a number of sectors.

[1 2] National Conference on Agricultural Mechanization, NCNA, Jan. 4, 1978; BBC FE/5707/B11/7.

Therefore, long-term scientific research requires more resources. On this matter there has been a clash with the supporters of alternative views propagated by the Gang of Four. They were saying that the research institutes should carry out open-door research in order to meet immediate needs in production.

Their views on research are, of course, related to their attitudes to researchers and other intellectuals. In the opinion of the Gang of Four, researchers must not be treated differently from ordinary people. Laymen and researchers should be equal. On that ground it was unacceptable that researchers could withdraw from the demands for manual labour, political work, etc., that other groups in Chinese society were requested to fulfil. It has now been clearly pointed out that such a situation hindered the research and technological development that China needed for the future. This problem was compounded by the fact that the students trained at the universities since the Cultural Revolution were insufficient in numbers and quality.

However, the more fundamental differences between the Gang of Four and the new Chinese leadership are more complex. The Cultural Revolution awakened strong social forces in Chinese society. Included here are the young people who in China, like many other developing countries, constitute 45% of the age groups under 20 years. The demand for a more equal and just society was only partially articulated in programmes which had any chance of implementation. The just society of which many had a vision should not accept any privileged groups. This criticism turned against the established institutions like universities and research institutes, and the class struggle was one of the instruments for achieving change. But the class struggle as a political instrument also led to an alternative view of the intellectuals. They were seen more or less as potential and permanent enemies, and therefore the class struggle within universities and research institutes had to be institutionalized — according to the alternative view of the Gang of Four. The resulting political struggle in the institutes and the general demands for open-door research, no doubt, seriously lowered efficiency and quality.

The new party leadership has now resolved the issues and clearly stated that the alternative views were mistaken ones and would hinder the socialist transformation of Chinese society because they obstructed the development of the material base. The present view is that science and technology are a productive force — that is to say an instrument. Consequently, science and technology — and the scientists and engineers as well — should not be seen as part of the superstructure and no longer constitute a conflict with the economic base. Thus scientists and engineers should not be struggled against.

Another important consequence is that the organizational problems should be solved differently. If science and technology are a productive force, i.e. an instrument, the sector should be organized as efficiently as possible to make as early and as large a contribution as possible to the socialist transformation. Therefore, the scientists in the research institutes, who were previously considered as unreliable, need no longer be controlled within the groups to which they belong. External political control on how resources are utilized does not, of course, disappear. But the resources within the science and technology sector are seen as a key instrument for achieving long-term goals in China.

When we try to assess the changes in China it may always be useful to maintain a historical perspective as all changes may not be permanent, and we can expect that China will experience further political struggles between opposing views on the role of science and technology and how the sector should be controlled and organized. Is there any risk that China will eventually move towards political changes such as have taken place in the

Soviet Union and which the Chinese term revisionism? No doubt, the heavy emphasis on economic growth and the use of intellectual and technological expertise in the country may make it difficult to strike a correct balance. An awareness of this problem is obvious from the following comment in an article published in 1977 after the fall of the Gang of Four:

> True, in the Soviet Union, the red flag has fallen to the ground, but that is because the Soviet revisionist renegade clique has usurped the supreme leadership of the Party and State and pushed a counter-revolutionary revisionist line. So long as we adhere to Chairman Mao's revolutionary line, persist in taking the socialist road and criticize revisionism and capitalism, we shall be able to keep the red flag flying, send our satellites into orbit, and build our country into a powerful modern socialist state.[13]

Nonetheless, the new technology and science policy now emerging in China may be an element which is at least partly antagonistic to the objective of reaching a true socialist society, and the reasons for this are several. First, to meet the technology requirements of modern industry, the emphasis will be on large-scale systems with a high degree of vertical division of labour and apparent consequences for management in enterprises as well as in the required R & D undertakings. Second, this is further underlined by importation of technology where technological and management solutions developed in capitalist countries cannot easily be changed to suit Chinese conditions, if this were desired. The integration and co-ordination of large-scale technological projects and the subsequent application in manufacturing will require expertise which must be highly trained and competent. All such people will spend much of their time in central agencies, ministries, or offices in the bigger cities with little, or at least less, time than previously to move into manual labour. Fourth, a large-scale approach to industrialization also requires improved transportation and communications and new management systems, which all lend some credibility to the argument that new forms of social control might develop which are detrimental to the interests of the masses of the Chinese population.

In all fairness, it must be pointed out that the policy statements so far available to us clearly indicate that the leadership is bent on maintaining a balance of technological development between the various sectors of the society — that is the continuation of a two-leg policy. However, in carrying out such a policy it is emphasized that the professionals are to form the backbone in any undertakings where the masses are involved. This may gradually lead to a situation where indigenous technological development is being downplayed thereby favouring an increased emphasis on the importation of technology. More ambiguous effects could be found in a situation where the present momentum to reinstate the professionals — technicians, engineers, researchers, and other intellectuals — cannot be stopped. The result might then be that these people establish themselves as new privileged groups with the blessing of the party and military bureaucracy. Such a possibility cannot be ruled out because — if the food supply problem is being solved — there is no urgent need, in the short run, to divert large investment resources into agriculture, as this would only aggravate the issue of employment and labour allocation. For the time being China, like many other developing countries, has no other choice but to use agriculture as an employer for the majority of the yearly large increases in the labour force. Consequently, development resources such as engineering manpower, R & D resources, etc., are likely to flow mainly into the modern industrial sector. This might then add

---

[13] Chi Wei, "How the Gang of Four opposed socialist modernization", *Beijing Review*, 1977, No. 11.

additional support to the hypothetical possibility that certain key groups in the urban-based modern economy might establish themselves as privileged groups to the detriment of the majority of the population residing in rural areas.

So it might be appropriate to pose the following question. Will it be possible for the Chinese leadership to maintain a fair balance between industry and agriculture or rather between the urban and rural groups which boils down to the question of the attitudes of those making decisions over resources? Here we can find three different interpretations with regard to the present changes in technology and science policy. First, the leadership will be able to maintain the delicate but necessary balance in meeting the objectives reflecting the requirements of the various groups in Chinese society. Second, the potentially privileged groups are making use of the new situation in order to further their interests. Such a situation will, in particular, be detrimental to the majority of the population in the rural areas who, together with other not yet clearly identified groups, might create a counterforce in order to achieve their objectives. Should this be the case the present change in technology policy has apparently created an unstable situation. Third, the changes might create a situation where privileged groups become established as a new class and the situation may at least temporarily be stabilized – but to the detriment of the overall socialist development of the country.

From the confident Chinese statements on science and technology transpires an optimism and freshness which can no longer be found in already industrialized countries. In our countries we have, with few exceptions, a very gloomy view of what science and technology can do. We have misgivings about a complete computerization of our societies, deterioration of the environment, safety, and supply of energy production, etc. Still, the industrialized world has a virtual world monopoly on R & D resources. The developing countries want to break this monopoly as quickly as possible and join the club of the technologically and industrially privileged, and China is likely to be in the forefront of this movement.

There are a number of reasons for China's optimistic attitudes towards science and technology. The most important is the realization that technology is required to modernize and that there is no real substitute for modern technology and scientific research. Second, researchers, technicians, and other intellectuals were abused during the Cultural Revolution and now they are asked to come out and make their rightful contribution to the development of the country, and most of them are now more than happy to do this. Third, and not least important for the lack of gloomy views in China, is the possibility that many of the benefits and disadvantages of technological development can be better handled in China.

The consequence would then be that the same drawbacks such as alienation and industrial dislocation now plaguing industrialized countries are at least for the time being not apparent in China or much less so. This may be due among other things to a lower level of economic development and a very different social organization. Finally, we should remember that our knowledge of Chinese developments is still very fragmentary and often superficial, and we cannot really say that China is going to avoid all the evils of modernization and technological development. The debate on science and technology as reflected in the news media over the past couple of years can so far shed only a little light on the development of science and technology in China. For a fuller understanding it is necessary to follow actively and interpret continually developments in China and to try to relate achievements and changes both to the Chinese system and to the global situation.

# Appendix I.
# Selected Foreign Statistics on the People's Republic of China[a]

TABLE I.1. *Key Economic Indicators*

|  | 1952 | 1957 | 1965 | 1970 | 1975 | 1976 | 1977 |
|---|---|---|---|---|---|---|---|
| GNP (billion 1976 $US) | 87 | 122 | 165 | 231 | 323 | 324 | |
| Population (mid-year (million persons)) | 570 | 640 | 750 | 840 | 935 | 951 | |
| *Per capita* GNP (1976 $US) | 153 | 190 | 220 | 275 | 346 | 340 | |
| *Agricultural Production Index* (1957=100) | 83 | 100 | 104 | 127 | 148 | 148 | |
| Total grain (million tonnes) | 161 | 191 | 194 | 243 | 284 | 285 | |
| Cotton (million tonnes) | 1.3 | 1.6 | 1.9 | 2.0 | 2.3 | 2.3 | |
| Hogs (million head) | 58 | 115 | 168 | 226 | – | 280 | |
| *Industrial Production Index* (1957=100) | 48 | 100 | 199 | 316 | 502 | 502 | |
| *Producer Goods Index* (1957=100) | 39 | 100 | 211 | 350 | 602 | – | |
| *Machinery Index* (1957=100) | 33 | 100 | 257 | 586 | 1,156 | – | |
| Electric generators (million kW) | Negl | 0.3 | 0.8 | – | 5.5 | – | |
| Machine tools (thousand units) | 13.7 | 28.3 | 45.0 | 70.0 | 90.0 | – | |
| Tractors (thousand 15-hp units) | 0 | 0 | 23.9 | 79.0 | 180.0 | 190.9 | |
| Trucks (thousand units) | 0 | 7.5 | 30.0 | 70.0 | 133.0 | – | |
| Locomotives (units) | 20 | 167 | 50 | 435 | 530 | – | |
| Freight cars (thousand units) | 5.8 | 7.3 | 6.6 | 12.0 | 18.5 | – | |
| Merchant ships (thousand tonnes) | 6.1 | 46.4 | 50.6 | 121.5 | 313.6 | 318.8 | |
| *Other Producer Goods Index* (1957=100) | 41 | 100 | 200 | 294 | 472 | – | |
| Electric power (billion kWh) | 7.3 | 19.3 | 42.0 | 72.0 | 121.0 | – | |
| Coal (million tonnes) | 66.5 | 130.7 | 220.0 | 310.0 | 427.0 | 448.0 | |
| Crude oil (million tonnes) | 0.4 | 1.5 | 11.0 | 28.2 | 74.3 | 83.6 | |
| Crude steel (million tonnes) | 1.3 | 5.4 | 12.5 | 17.8 | 26.0 | 23.0 | |
| Chemical fertilizer (million tonnes) | 0.2 | 0.8 | 7.6 | 14.0 | 27.9 | – | |
| Cement (million tonnes) | 2.9 | 6.9 | 16.3 | 26.6 | 47.1 | 49.3 | |
| Timber (million m³) | 11.2 | 27.9 | 27.2 | 29.9 | 36.2 | – | |
| Paper (million tonnes) | 0.6 | 1.2 | 3.6 | 5.0 | 6.9 | – | |
| *Consumer Goods Index* (1957=100) | 60 | 100 | 183 | 272 | 368 | – | |
| Cotton cloth (billion linear metres) | 3.8 | 5.0 | 6.4 | 7.5 | 7.6 | – | |
| Wool cloth (million linear metres) | 4.2 | 18.2 | – | – | – | – | |
| Processed sugar (million tonnes) | 0.5 | 0.9 | 1.5 | 1.8 | 2.3 | – | |
| Bicycles (million units) | 0.1 | 0.8 | 1.8 | 3.8 | 5.5 | – | |
| *Foreign Trade* (billion current $US) | 1.9 | 3.0 | 3.8 | 4.3 | 14.4 | 12.9 | |
| Exports, f.o.b. | 0.9 | 1.6 | 2.0 | 2.0 | 7.0 | 6.9 | 7.8 |
| Imports, c.i.f. | 1.0 | 1.4 | 1.8 | 2.2 | 7.4 | 6.0 | 6.4 |

[a]At the time of writing, aggregate statistical information was not available from Chinese official sources. Faced with the dilemma of either using foreign estimates, which may occasionally not be very reliable or not providing the reader with any substantial statistics at all, the decision was reached to use estimates published by the US Government. The tables in this appendix are from *China: Economic Indicators – A Reference Aid*, National Foreign Assessment Center, Washington DC, Oct. 1977.

TABLE I.2. *Indicators of Aggregate Performance*

| | Gross national product (billion 1976 $US) | Index: 1957=100 | |
| | | Agricultural production | Industrial production |
| --- | --- | --- | --- |
| 1949 | 51 | 54 | 20 |
| 1950 | 63 | 64 | 27 |
| 1951 | 74 | 71 | 38 |
| 1952 | 87 | 83 | 48 |
| 1953 | 93 | 83 | 61 |
| 1954 | 97 | 84 | 70 |
| 1955 | 106 | 94 | 73 |
| 1956 | 115 | 97 | 88 |
| 1957 | 122 | 100 | 100 |
| 1958 | 145 | 108 | 142 |
| 1959 | 138 | 84 | 173 |
| 1960 | 134 | 74 | 181 |
| 1961 | 106 | 79 | 105 |
| 1962 | 118 | 89 | 111 |
| 1963 | 132 | 96 | 134 |
| 1964 | 149 | 103 | 161 |
| 1965 | 165 | 104 | 199 |
| 1966 | 185 | 113 | 232 |
| 1967 | 178 | 118 | 202 |
| 1968 | 179 | 110 | 221 |
| 1969 | 199 | 113 | 266 |
| 1970 | 231 | 127 | 316 |
| 1971 | 247 | 130 | 349 |
| 1972 | 258 | 126 | 385 |
| 1973 | 292 | 142 | 436 |
| 1974 | 302 | 146 | 455 |
| 1975 | 323 | 148 | 502 |
| 1976 | 324 | 148 | 502 |

TABLE I.3. *Indicators of Aggregate Performance*

| | Index: 1957=100 | | Foreign trade (billion current $US) | |
|---|---|---|---|---|
| | Construction activity | Modern transport performance | Exports (f.o.b.) | Imports (c.i.f.) |
| 1949 | 13 | 16 | 0.4 | 0.4 |
| 1950 | 21 | 27 | 0.6 | 0.6 |
| 1951 | 32 | 31 | 0.8 | 1.1 |
| 1952 | 41 | 39 | 0.9 | 1.0 |
| 1953 | 53 | 49 | 1.0 | 1.3 |
| 1954 | 72 | 62 | 1.1 | 1.3 |
| 1955 | 73 | 65 | 1.4 | 1.7 |
| 1956 | 110 | 87 | 1.6 | 1.5 |
| 1957 | 100 | 100 | 1.6 | 1.4 |
| 1958 | 149 | 148 | 1.9 | 1.8 |
| 1959 | 173 | 201 | 2.2 | 2.1 |
| 1960 | 161 | 196 | 2.0 | 2.0 |
| 1961 | 102 | 132 | 1.5 | 1.5 |
| 1962 | 92 | 136 | 1.5 | 1.2 |
| 1963 | 116 | 148 | 1.6 | 1.2 |
| 1964 | 140 | 156 | 1.8 | 1.5 |
| 1965 | 182 | 172 | 2.0 | 1.8 |
| 1966 | 197 | 192 | 2.2 | 2.0 |
| 1967 | 157 | 161 | 1.9 | 2.0 |
| 1968 | 202 | 170 | 1.9 | 1.8 |
| 1969 | 230 | 203 | 2.0 | 1.8 |
| 1970 | 266 | 245 | 2.0 | 2.2 |
| 1971 | 300 | 286 | 2.4 | 2.3 |
| 1972 | 351 | 302 | 3.1 | 2.8 |
| 1973 | 369 | 326 | 5.0 | 5.1 |
| 1974 | 335 | 340 | 6.6 | 7.4 |
| 1975 | 404 | 372 | 7.0 | 7.4 |
| 1976 | – | – | 6.9 | 6.0 |
| 1977 | – | – | 7.8 | 6.4 |

TABLE I.4. *Estimated Population*[(a)] *by Age and Sex (July 1, 1977)*

|  | Million persons | | | Per cent distribution | | |
|---|---|---|---|---|---|---|
|  | Total | Males | Females | Total | Males | Females |
| All ages | 966 | 486 | 480 | 100 | 100 | 100 |
| 0–4 | 122 | 62 | 60 | 12.58 | 12.72 | 12.45 |
| 5–9 | 128 | 65 | 63 | 13.26 | 13.37 | 13.15 |
| 10–14 | 112 | 57 | 55 | 11.59 | 11.66 | 11.52 |
| 15–19 | 99 | 50 | 49 | 10.30 | 10.32 | 10.27 |
| 20–24 | 96 | 48 | 47 | 9.89 | 9.92 | 9.86 |
| 25–29 | 77 | 39 | 38 | 8.00 | 8.07 | 7.94 |
| 30–34 | 60 | 31 | 29 | 6.23 | 6.35 | 6.12 |
| 35–39 | 53 | 27 | 26 | 5.46 | 5.56 | 5.35 |
| 40–44 | 49 | 25 | 24 | 5.10 | 5.17 | 5.02 |
| 45–49 | 43 | 21 | 21 | 4.42 | 4.40 | 4.44 |
| 50–54 | 36 | 17 | 18 | 3.68 | 3.60 | 3.76 |
| 55–59 | 29 | 14 | 15 | 2.99 | 2.85 | 3.13 |
| 60–64 | 23 | 11 | 12 | 2.39 | 2.25 | 2.52 |
| 65–69 | 18 | 8 | 9 | 1.81 | 1.70 | 1.93 |
| 70–74 | 12 | 5 | 6 | 1.23 | 1.13 | 1.34 |
| 75 and over | 10 | 5 | 6 | 1.07 | 0.93 | 1.21 |

[(a)]These estimates were prepared by the US Department of Commerce, Bureau of Economic Analysis, Foreign Demographic Analysis Division.

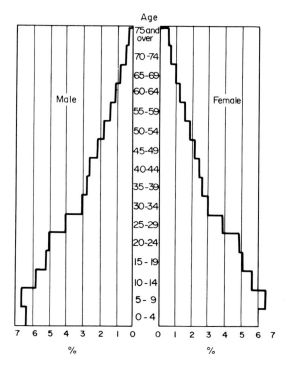

FIG. I.1. *China: estimated age–sex distribution, July 1, 1977*

# Appendix II.
# China's Six Economic Regions[1]

AT THE National Conference, following the Daqing example in industry, convened in April and May 1977, Vice-premier Yu Qiuli put forth as one of China's future goals the policy of gradually establishing an economic system involving relatively balanced development of agriculture, light industries, and heavy industries in the six large regions Dongbei (north-east), Huabei (north), Hua-dong (east), Zhongnan (central-south), Xinan (south-west), and Xibei (north-west), so that they can function independently and can extensively co-operate with one another with their different levels of development and respective characteristics. The regions are delineated in Fig. II.1.

Total area: 9.6 million km².

FIG. II.1.

A. North China (Huabei): Beijing, Tianjin, Hebei, Shanxi, Nei Monggol (Inner Mongolia)
B. North-east (Dongbei): Liaoning, Jilin, Heilongjiang
C. North-west (Xibei): Shaanxi, Gansu, Qinghai, Xinjiang
D. East China (Huadong): Shanghai, Shandong, Jiangsu, Anhui, Zhejiang, Fujian, Jiangxi
E. Central-south (Zhongnan): Henan, Hubei, Hunan, Guangdong, Guangxi
F. South-west (Xinan): Sichuan, Guizhou, Yunnan, Xizang (Tibet)

China was divided into great administrative regions until 1954. The concept of economic co-ordinating regions – seven at the time – first started to be mentioned in 1958. However, it is unclear if any economic planning based on such regions was ever implemented.[2] JETRO has attempted to estimate various economic parameters of the newly announced regions, illustrated in Fig. II.2.

[1] From: *JETRO China Newsletter,* No. 17 (Apr. 1978) pp. 23–24.
[2] The interested reader might consult A. Donnithorne, *China's Economic System,* George Allen & Unwin, London, 1967.

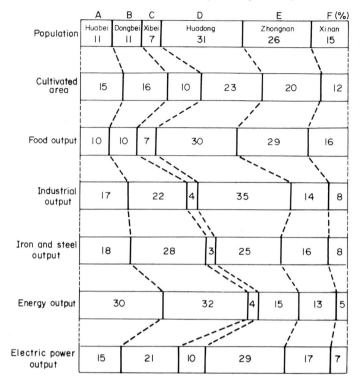

FIG. II.2. *The economic composition of the six regions of China.*

# Appendix III.
# The State Council[1] – Members and Agencies

HUA GUOFENG, *Premier of the State Council*
DENG XIAOPING, *Vice-premier*
LI XIANNIAN, *Vice-premier*
XU XIANGQIAN, *Vice-premier*
JI DENGGUI, *Vice-premier*
YU QIULI, *Vice-premier*
CHEN XILIAN, *Vice-premier*

GENG BIAO, *Vice-premier*
CHEN YONGGUI, *Vice-premier*
FANG YI, *Vice-premier*
WANG ZHEN, *Vice-premier*
GU MU, *Vice-premier*
KANG SHIEN, *Vice-premier*
CHEN MUHUA, *Vice-premier*

| *Minister* | | *Ministry/Agency* |
|---|---|---|
| HUANG HUA | | Foreign Affairs |
| XU XIANGJIAN | (in charge of) | National Defence |
| YU QIULI | (in charge of) | State Planning Commission |
| KANG SHIEN | (in charge of) | State Economic Commission |
| GU MU | (in charge of) | State Capital Construction Commission |
| FANG YI | (in charge of) | State Scientific and Technological Commission |
| YANG JINGREN | (in charge of) | Nationalities Affairs Commission |
| CHAO CANGBI | | Public Security |
| CHENG ZIHUA | | Civil Affairs |
| LI QIANG | | Foreign Trade |
| CHEN MUHUA | | Economic Relations with Foreign Countries |
| YANG LIGONG | | Agriculture and Forestry |
| TANG KE | | Metallurgical Industry |
| CHOU ZIJAN | | First Ministry of Machine-building |
| LIU WEI | | Second Ministry of Machine-building |
| LÜ DONG | | Third Ministry of Machine-building |
| WANG ZHENG | | Fourth Ministry of Machine-building |
| ZHUNG ZHEN | | Fifth Ministry of Machine-building |
| CHAI SHUFAN | | Sixth Ministry of Machine-building |
| SONG RENQIONG | | Seventh Ministry of Machine-building |
| XIAO HAN | | Coal Industry |
| SONG ZHENMING | | Petroleum Industry |
| SUN JINGWEN | | Chemical Industry |
| QIAN ZHENGYING | | Water Conservancy and Power |
| QIAN LINGGUANG | | Textile Industry |
| LIANG LINGGUANG | | Light Industry |
| DUAN JUNYI | | Railways |
| YE FEI | | Communications |
| ZHONG FUXIANG | | Post and Telecommunications |
| ZHANG JINGFU | | Finance |
| LI BAOHUA | (*President*) | People's Bank of China |
| WANG LEI | | Commerce |
| ZHEN GUODONG | (*Director*) | All-China Federation of Supply and Marketing Cooperatives |
| HUANG ZHEN | | Culture |
| LIU XIYAO | | Education |
| JIANG YIZHEN | | Public Health |
| WANG MENG | (in charge of) | State Physical Culture and Sports Commission |

[1] As reported from the Fifth National People's Congress by the New China News Agency, Mar. 5, 1978.

# Appendix IV.

# Research Institutes under the Chinese Academy of Sciences[1] and their Main Research Activities (June 1978)[2]

MATHEMATICS, PHYSICS, AND CHEMISTRY

1. *Institute of Mathematics (Beijing)*
   Pure and applied mathematics.

2. *Institute of Mechanics (Beijing)*
   Solid-body mechanics, fracture mechanics, electromagnetic hydro-mechanics, fluid mechanics, explosion mechanics, thermophysics (including power thermophysics and gas kinetics, heat transfer, motors), and related subjects.

3. *Hubei Institute of Rock and Soil Mechanics (Wuhan)*
   Mechanics of rock, soil, and related media.

4. *Institute of Physics (Beijing)*
   Solid-state physics, including research work in low-temperature superconductivity, high pressure, crystallography, magnetism, lasers, and other areas. Experimental research in plasma physics and in small-scale controlled thermonuclear reactors.

5. *Institute of Atomic Energy (Beijing)*
   Nuclear physics, radiochemistry, reactor physics, nuclear electronics, isotope production, etc.

6. *Institute of High-energy Physics (Beijing)*
   Structure of elementary particles and their laws of motion and transformation. Preparation of the experimental basis for the construction of a high-energy accelerator.

7. *Yunnan Branch Institute of Cosmic Rays (Kunming)*
   High-energy physics by searching for new particles and novel high-energy phenomena in cosmic rays.

8. *Institute of Modern Physics (Lanzhou, Gansu)*
   Experimental work and theoretical research in heavy-ion nuclear physics. Preparation for the construction of a heavy-ion accelerator.

---

[1] The newly formed Chinese Academy of Social Sciences, which partly corresponds to the previous Department of Social Sciences of the Chinese Academy of Sciences, exercises leadership over the following research institutes

| | |
|---|---|
| Institute of Literature | Institute of Information |
| Institute of Foreign Literature | Institute of Economics |
| Institute of Linguistics | Institute of World Economics |
| Institute of History | Institute of Industrial Economics |
| Institute of Modern History | Institute of Agricultural Economics |
| Institute of World History | Institute of Finance and Trade |
| Institute of Philosophy | Institute of Technological Economics |
| Institute of World Religions | Institute of Journalism |
| Institute of Law | Institute of South Asian Studies |
| Institute of Archaeology | Institute of USA (planned) |
| Institute of Nationalities | Institute of USSR (planned) |
| Institute of Japan (planned) | Institute of South East Asia (planned) |
| Institute of West Europe (planned) | Institute of Sociology (planned) |

[2] The information here is based on an official Chinese list supplied in August 1978 and includes only those institutes which were at the time under Academy leadership. I am indebted to Donald Blackmore Wagner, research fellow at the Scandinavian Institute of Asian Studies, for making a competent translation, at short notice, of the above list.

9. *Shanghai Institute of Nuclear Physics*
   Theoretical and applied research in low- and intermediate-energy nuclear physics. Isotope applications.

10. *Institute of Acoustics (Beijing)*
    Underwater sound, ultrasonics, electro-acoustics, noise, language acoustics, etc.

11. *Jilin Institute of Physics (Changchun)*
    Luminescence of solids and display technology.

12. *Xinan Institute of Physics (Leshan, Sichuan)*
    Plasma physics and controlled thermonuclear reactions.

13. *Institute of Plasma Physics (Hefei, Anhui)*
    Preparatory construction for a large-scale controlled thermonuclear reactor; research in high-temperature plasma physics.

14. *Institute of Chemistry (Beijing)*
    Modern physical chemistry, high-polymer chemistry and physics (including reinforced plastics, plastic adhesives, and new types of elastic materials), and organic photoelectric materials.

15. *Institute of Photochemistry (Beijing)*
    Light-sensitive chemistry and photochemistry.

16. *Institute of Environmental Chemistry (Beijing)*
    Pollution chemistry and analytical chemistry and regulation.

17. *Institute of Chemical Metallurgy (Beijing)*
    Physical chemistry of metallurgical processes and chemical reaction engineering.

18. *Shanghai Institute of Organic Chemistry*
    Natural organic chemistry, element organic chemistry, and theoretical organic chemistry.

19. *Shanghai Institute of Silicate Chemistry and Engineering*
    New types of inorganic non-metallic materials, including crystals. Applied and theoretical research on ceramics and glass.

20. *Fujian Institute of Research in the Structure of Matter (Fuzhou)*
    Research in the structure of matter (including crystallography, catalysis, and chemical analogues of biological nitrogen fixation).

21. *Lanzhou (Gansu) Institute of Chemical Physics*
    Chemical catalysis, surface chemistry, solid lubricating materials, etc.

22. *Dairen (Jilin) Institute of Chemical Physics*
    Reaction kinetics of catalysts and catalytic processes, and related research in analytical chemistry.

23. *Jilin Institute of Applied Chemistry (Changchun)*
    High-polymer chemistry and physics, rare-earth chemistry, and related research in analytical chemistry.

24. *Guangdong Institute of Chemistry (Guangzhou)*
    High-polymer molding materials and the chemical modification of natural fibres.

GEOSCIENCES

25. *Beijing Observatory*
    Celestial physics and radio astronomy; also research in prediction of solar activity.

26. *Purple Mountain Observatory (Nanjing)*
    Celestial mechanics and radio astronomy; also research on the sun, the fixed stars, the planets, and ancient Chinese astronomy.

27. *Shanghai Observatory*
    Astrogeodesy, short-wave time service, etc.

28. *Yunnan Observatory (Kunming)*
    Stellar physics, solar physics, and radio astronomy.

29a. *Institute of Geography (Beijing)*
    China's geography, emphasizing research on the laws of formation and modification of the natural geographical environment; also questions concerning the transformation and exploitation of natural conditions and the development and exploitation of natural resources.

29b. *Institute of Geography, Section 2 (Beijing)*
      Applications of remote-sensing techniques and automatization of cartography.

30.  *Institute of Geology (Beijing)*
      Formation and structure of the lithosphere and the fundamental laws of geological evolution; investigation of the history of geological evolution in China and questions of global structure.

31.  *Institute of Geophysics (Beijing)*
      Internal structure, state, composition, and motion of the earth, and its internal physical processes.

32.  *Institute of Atmospheric Physics (Beijing)*
      Fundamental laws of atmospheric motion and regularities in important physical phenomena of the atmosphere. Conducts research work in atmospheric kinetics, cloud physics, turbulence diffusion, atmospheric surveying, etc.

33.  *Lanzhou (Gansu) Institute of Plateau Atmospheric Physics*
      Fundamental laws of atmospheric motion in the northwestern plateau area, and in regularities in important physical phenomena of the atmosphere.

34.  *Guiyang (Guizhou) Institute of Geochemistry*
      Basic principles of metallogenetics, diagenetics, and mineralogy. Also concerned with research work in geohistory, mineral physics, experimental geology, and the geology of the Quaternary period.

35.  *Institute of Vertebrate Palaeontology and Palaeanthropology (Beijing)*
      Formal classification and phylogenesis of palaeovertebrates and related theoretical topics in biological history. Geographical and chronological distribution of Chinese fossil vertebrates; the origin and evaluation of the anthropoids and the higher primates, and the formation of modern man; palaeolithic culture; sites of primitive man and their environment.

36.  *Nanjing Institute of Geopalaeontology*
      Relations between stratigraphy and palaeontology in order to provide support for the search for iron, coal, oil, and other necessary mineral resources; investigations of the relations between accumulations (especially sedimentary environments) and biological forms (e.g. palaeobiota), and the history of biological change and development.

37.  *Lanzhou (Gansu) Institute of Glaciers and Frozen Soil*
      Fundamental theoretical problems related to glaciers and frozen soil; co-ordination of construction engineering work related to mining, transportation, etc.; research on problems of control of avalanches (of ice and snow, frozen earth, mud, and rocks) and other natural disasters.

38.  *Lanzhou (Gansu) Institute of Deserts*
      Formation, transformation, and reclamation of deserts.

39.  *Institute of Oceanology (Qingdao, Shandong)*
      General investigations and basic research in the hydrography, climatology, chemistry, geology, geophysics, and marine biology of Chinese coastal regions and the neighbouring ocean areas.

40.  *South China Sea Institute of Oceanology (Guangzhou)*
      General exploration of the islands of the South China Sea and the adjacent ocean regions; research in marine geology, ocean pollution, and marine aquaculture.

41.  *Committee for General Investigation of Natural Resources (Beijing)*
      Organization and Co-ordination of General Investigations concerning Chinese natural resources (especially agricultural resources). Carries out general analyses of the results of some of these investigations and presents views concerning development and conservation.

BIOSCIENCES

42.  *Institute of Zoology (Beijing)*
      Research in zoology and entomology. Conducts applied research in insect-pest control and some basic research, e.g. mediation mechanisms of nucleic acid.

43.  *Yunnan Institute of Biology (Kunming)*
      Research in systematic zoology, zoological identification, animal ecology, cell biology, etc.

44.  *Shanghai Institute of Entomology*
      Systematic entomology, insect ecology, and the control of insect–pest diseases in agriculture.

45. *Institute of Microbiology (Beijing)*
    Classification and bioculture preservation of all forms of micro-organisms; exploitation of China's microbiological resources. Also some representative work on metabolic control of micro-organisms.

46. *Institute of Psychology (Beijing)*
    Physiological psychology and engineering (applied?) psychology.

47. *Shanghai Institute of Physiology*
    Neurophysiology, physiology of senses, principles of acupuncture anaesthesia, and engineering noise control.

48. *Institute of Biophysics (Beijing)*
    Radiobiology, reproduction, and instrumentation; also research in fundamental aspects of insulin.

49. *Shanghai Institute of Biochemistry*
    Molecules on biology, including proteins, nucleid acids, enzymes, etc.

50. *Shanghai Institute of Cell Biology*
    Cell biology, tumours, etc.

51. *Shanghai Institute of Materia Medica*
    Pharmacology, pharmaceutical chemistry, drug synthesis, and anti-tumour drugs.

52. *Institute of Genetics (Beijing)*
    Laws of hereditary change in sexual and asexual processes, and their exploitation in breeding; the influence of environment on character formation and development; and the material basis of genetics.

53. *Institute of Botany (Beijing)*
    Systematic botany and the development and utilization of botanical resources; development of new opportunities for agriculture.

54. *Guangdong Institute of Botany (Guangzhou)*
    Systematic botany, plant ecology, geobotany, plant physiology, cell botany, plant genetics, plant domestication, etc., of tropical Asia.

55. *Yunnan Institute of Botany (Kunming)*
    Systematic botany, plant ecology, geobotany, plant physiology, botanochemistry, plant domestication, etc., of south-west China.

56. *Shanghai Institute of Plant Physiology*
    Photosynthetic functions, plant cell physiology, plant nourishment physiology, genetic engineering, and botanical nitrogen fixation.

57. *Nanjing Institute of Pedology*
    Evaluation of China's soil resources and their optimum utilization; also work in soil conservation, methods of soil measurement, etc.

58. *Institute of Forestry and Pedology (Shenyang, Liaoning)*
    Pedology, botany, microbiology, forest ecology, etc., of Manchuria.

NEW TECHNIQUES

59. *Institute of Computing Technology (Beijing)*
    Theory and techniques of computer systems (hardware and software); large-scale computer systems and networks.

60. *Shenyang (Liaoning) Institute of Computer Technology*
    Development and production of small- and medium-scale general-purpose computers; hardware research.

61. *Institute of Automation (Beijing)*
    Theoretical and applied research on automatic control; research on control mechanisms and information control systems; character and image recognition.

62. *Shenyang (Liaoning) Institute of Automation*
    Research on control systems. In addition, research on servomechanisms and information control systems, and on control theory and new automation techniques.

63. *Institute of Electrical Engineering (Beijing)*
    New techniques in electrical engineering and new applications of electrical energy, e.g. magnetic fluid generators, superconducting magnets, etc.

64. *Institute of Electronics (Beijing)*
    Microwave electronics, quantum electronics, photoelectric techniques, cathode electronics, electron optics, etc.

65. *Shanghai Institute of Technical Physics*
    Infrared technology; also infrared physics.

66. *Changchun (Jilin) Institute of Optical and Precision Instruments*
    Optical materials, optical apparatus, precision instrument technology, spectroscopy, etc.

67. *Shanghai Institute of Optical and Precision Instruments*
    Lasers and their applications; also research on laser theory.

68. *Xian (Shaanxi) Institute of Optical and Precision Instruments*
    Theoretical and applied research on high-speed cameras; also fibre optics and electron optics.

69. *Institute of Semiconductors (Beijing)*
    Semiconductor physics and application fundamentals; also on new semiconductor materials, new devices, new techniques and new circuits.

70. *Shanghai Institute of Metallurgy*
    Research on new semiconductor materials, new devices, and new techniques; also superconducting materials, magnetic materials, metal corrosion, etc.

71. *Institute of Metals (Shenyang, Liaoning)*
    Research in materials science and the physics of metals.

72. *Institute of History of Natural Sciences (Beijing)*
    Research on the history of the development of modern natural science.

# Appendix V.
# Selected Scientific and Technical Journals[1]

| Name | Number of issues per year | Publisher |
|---|---|---|
| *Acta Archaeologica Sinica*<br>Kaogu Xuebao | 2 | Archaeological Institute, Chinese<br>Academy of Social Sciences |
| *Acta Astronomica Sinica*<br>Tienwen Xuebao | 2 | China Astronomical Society |
| *Acta Botanica Sinica*<br>Zhiwu Xuebao | 4 | China Botanical Society |
| *Acta Biochimica et Biophysica Sinica*<br>Shengwu Huaxue yu Shengwu wuli<br>    Xuebao | 4 | |
| *Acta Chimica Sinica*<br>Huaxue Xuebao | 4 | China Chemical Society |
| *Acta Entomologica Sinica*<br>Kunchong Xuebao | 4 | China Entomological Society |
| *Acta Genetica Sinica*<br>Yichuan Xuebao | 4 | |
| *Acta Geologica Sinica*<br>Dizhi Xuebao | 4 | China Geological Society |
| *Acta Geophysica Sinica*<br>Diqiu Xuebao | 4 | China Geophysical Society |
| *Acta Mathematica Sinica*<br>Shuxue Xuebao | 4 | |
| *Acta Mechanica Sinica*<br>Lixue Xuebao | 4 | China Mechanics Society |
| *Acta Microbiologica Sinica*<br>Weishengwu Zazhi | 4 | China Microbiological Society |
| *Acta Palaeontologica Sinica*<br>Gushengwu Xuebao | 4 | China Palaeontological Society |

[1] The information on which this list is based was generously provided by the Office of the Scientific and Technical Attache at the Swedish Embassy in Beijing. It should be emphasized that the number of scientific and technical journals being made available for official distribution is constantly increasing. For example, it was announced in mid-1978 that the following eleven medical journals will be published:

*The Chinese Journal of Preventive Medicine*
*The Chinese Journal of Obstetrics and Gynaecology*
*The Chinese Journal of Paediatrics*
*The Chinese Journal of Radiology*
*The Chinese Journal of Ophthalmology*
*The Chinese Journal of Otorhinolaryngology*
*The Chinese Journal of Stomatology*
*The Chinese Journal of Neuropsychiatry*
*The Chinese Journal of Laboratory Diagnosis*
*Progress in Physiological Sciences*
*The Chinese Journal of Tuberculosis and Respiratory Diseases*

| Name | Number of issues per year | Publisher |
|------|:---:|------|
| *Acta Physica Sinica*<br>Wuli Xuebao | 6 | China Physics Society |
| *Acta Phytotaxonomica Sinica*<br>Zhiwufenlei Xuebao | 4 | China Botanics Society |
| *Acta Zoologica Sinica*<br>Dongwu Xuebao | | |
| *Aeronautical Knowledge*<br>Hangkong Zhishi | 12 | China Aeronautical Society |
| *Agricultural Machines*<br>Nongye Jixie | 12 | Agricultural Machinery Scientific Research Institute, First Ministry of Machine-building |
| *Archaeology*<br>Kaogu | 6 | |
| *Architectural Journal*<br>Jianxhu Xuebao | 4 | China Architectural Society |
| *Botanical Journal*<br>Zhiwu Zazhi | 6 | |
| *Bulletin of Microbiology*<br>Weishengwuxue Tongbao | 6 | China Society of Microbiology |
| *Chemical Bulletin*<br>Huaxue Tongbao | 6 | China Chemical Society |
| *Chinese Forestry Science*<br>Zhongguo Linye Kexue | 4 | Chinese Academy of Forestry |
| *Chinese Journal of Internal Medicine*<br>Zhonghua Neike Zazhi | 6 | Chinese Medical Association |
| *Chinese Medical Journal* (in English) | 6 | Chinese Medical Association |
| *Cultural Relics*<br>Wenwu | 12 | |
| *Development of Biochemistry and Biophysics*<br>Shengwu Huaxue yu Shengwu Wuli Jinzhan | 6 | Institute of Biophysics, Academia Sinica |
| *Earthquake Front*<br>Dizhen Zhanxian | 6 | |
| *Economic Research*<br>Jingji Yanjiu | 12 | |
| *Electronics Science and Technology*<br>Dianzi Kexue Jishu | 12 | |
| *Entomological Knowledge*<br>Kunchong Zhishi | 6 | China Entomological Society |
| *Environmental Science*<br>Huanjing Kexue | 6 | Institute of Environmental Chemistry, Academia Sinica |
| *Fossil*<br>Huashi | 4 | |
| *Fudan Journal (Natural Science)*<br>Fudan Xuebao, Zirankexueban | 4 | |
| *Geochimica*<br>Diqiu Huaxue | 4 | Guyiang Institute of Geochemistry, Academia Sinica |

| Name | Number of issues per year | Publisher |
|---|---|---|
| *Geographical Knowledge* Dili Zhishi | 12 | |
| *Heredity and Breeding* Yichuan yu Yuzhong | 6 | Institute of Genetics, Academia Sinica |
| *High Energy Physics* Gaoneng Wuli | 4 | |
| *Journal of Tsinghua University* Qinghua Daxue Xuebao | 4 | |
| *Laser* Jiguang | 6 | |
| *Middle School Science and Technology* Zhongxue Keji | 6 | |
| *National Medical Journal of China* Zhonghua Yixue Zazhi | 12 | Chinese Medical Association |
| *Nature Journal* Ziran Zazhi | 12 | |
| *Oceanologica et Limnologica Sinica* Haiyang yu Huzhao | 4 | Chinese Society for Oceanology and Limnology |
| *Physica Energiae Fortis et Physica Nuclearis* Gaoneng Wuli yu Hewuli | 6 | |
| *Physics* Wuli | 6 | China Physics Society |
| *Practice and Knowledge of Mathematics* Shuxue de Shijian yu Renshi | 4 | |
| *Radio* Wuxiandian | 12 | |
| *Rural Scientific Experiments* Nonacun Kexue Shiyan | 12 | |
| *Scientia Agricultura Sinica* Zhingguo Nongye Kexue | 4 | Chinese Academy of Agricultural Sciences |
| *Scientia Atmospherica Sinica* Daqi Kexue | 4 | Institute of Atmospheric Physics, Academia Sinica |
| *Scientia Geologica Sinica* Dizhi Kexue | 4 | Geological Institute, Academia Sinica |
| *Scientia Sinica* Zhongguo Kexue | 6 | |
| *Science Bulletin* Kexue Tongbao | 12 | |
| *Science Pictorial* Kexue Huabao | 12 | |
| *Scientific Experiments* Kexue Shiyan | 12 | |
| *Vertebrata Palasiatica* Gujizhui Dongwu yu Gurenlei | 4 | Chinese Institute of Vertebrate Palaeontology Palaeanthropology, Academia Sinica |
| *Zoological Journal* Dongwuxue Zazhi | 4 | Chinese Zoological Society |

# Bibliography

*Acupuncture Anesthesia in the People's Republic of China,* a trip report of the American Acupuncture Anesthesia Study Group. Washington DC, National Academy of Sciences, 1976.

ANDERSSON SHIH, J., Science and technology in China, *Asian Survey,* Aug. 1972, pp. 662–75.

ANDORS, S., *China's Industrial Revolution – Politics, Planning and Management 1949 to the Present,* Pantheon Books, New York, 1977.

*Annotated Bibliography on Science and Technology in China,* US House of Representatives, Committee on Science and Technology. Washington DC, Government Printing Office, 1976, 52 pp.

AUDETTE, D. G., Computer technology in Communist China, 1956–1965, *Communications of the ACM,* Vol. 9 (1966), pp. 655–61.

BAARK, E., *Dissemination Structures for Technical Information in China – An Analysis of Three Industrial Sectors: Electronics, Metallurgy, and Agricultural Machinery,* interim report (Research Policy Programme), Lund, Aug. 1978.

BERBERET, J. A., *The Prospects for Chinese Science and Technology,* Tempo 68TMP-26 Feb. 1968, General Electric Co., Santa Barbara, Calif.

BERNER, B., *China's Science through Visitors' Eyes,* (Research Policy Programme), University of Lund, Sweden, June 1975, 58 pp.

BETTELHEIM, C., *Cultural Revolution and Industrial Organization in China. Changes in Management and the Division of Labor,* Monthly Reviewed Press, New York, 1974, 128 pp.

BILLGREN, B. and SIGURDSON, J., *An Estimate of Research and Development Expenditures in the People's Republic of China in 1973.* OECD Development Centre, Paris, July 1977, 45 pp.

BLAIR, P., *Development in the People's Republic of China: A Selected Bibliography,* Overseas Development Council, Dec. 1976.

BOXER, B. and PRAMER, D. (eds.), *Environmental Protection in the People's Republic of China,* Rutgers, the State University of New Jersey, 1978.

BUCK, P., *Order and Control: The Scientific Method in China and the United States, Social Studies of Science,* vol. 5, No. 3 (Aug. 1975), pp. 237–67.

BURCHETT, A. and ALLEY, R., *China: the Quality of Life,* Penguin Books, Harmondsworth, 1976, 312 pp.

CHANG, P. H., China's scientists in the Cultural Revolution, *Bulletin of the Atomic Scientists,* May 1969, pp. 10–20 and 40.

CHEATHAM, Jr., T. E., *et al.,* Computing in China: a travel report (reprint) *Science,* Vol. 182 (Oct 12, 1973), pp. 134–40.

CHENG, C. Y., *Scientific and Engineering Manpower in Communist China, 1949–63,* NSF, Washington, 1965.

CHIN, R. and CHIN, Ai-li, *Psychological Research in Communist China: 1949–1966,* Lexington, Mass., M I T Press, 1969.

CHINA, Recycling of organic wastes in agriculture, *FAO Soils Bulletin,* Rome, 1977.

CHINA, *Science Walks on Two Legs,* a report from Science for the People, Discus Books/Avon Books, New York, 1974, 316 pp.

CHINA and disarmament. World armaments and disarmaments, *Sipri Yearbook 1972,* Almqvist & Wiksell, Stockholm 1972, pp. 483–500.

CHINA's *developmental experience.* (Michael Oksenberg, ed.), Columbia University, the Academy of Political Science, New York, Mar. 1973, 227 pp.

*China's Road to Development,* 2nd edn. (N. Maxwell ed.), Pergamon Press, Oxford, 1979, 364 pp.

CHU, T. K., *Plasma Physics Laboratory. Current Plasma Physics Programs at Beijing Physics Institute,* Princeton, NJ, 1975 (TM-285), Princeton University, 8pp.

COE, R. S., Earthquake Prediction Program, *EOS,* Vol. 52, No. 12, (1971), pp. 940–3.

*Current Science,* Vol. XIV, No. 6, June 1976, PRC scientific organizations, pp. 15–21.

DEAN, G., China's Technological development, *New Scientist,* Vol. 54 (1972): 1976 (May 18), pp. 371–3.

DEAN, G., A note on the sources of technological innovation in the People's Republic of China, *Journal of Development Studies,* Vol. 9, No. 1 (Oct. 1972), pp. 187–99.

DEAN, G., *Science and Technology in the Development of Modern China. An annotated bibliography,* Mansell Information Publishing Ltd, London 1974, 265 pp.

DEAN, G., Science and the thought of Chairman Mao, *New Scientist,* Feb. 12, 1970, pp. 298–9.

DEAN, G., Science, technology and development: China as a case study, *The China Quarterly (London),* No. 51 (July–Sept. 1972), pp. 520–34.

DEAN, G. and MACIOTI, M., Scientific institutions in China, *Minerva,* Vol. XI, No. 3 (July 1973), pp. 318–34.

DEDIJER, S. and BILLGREN, B., The east is read, *Nature,* Vol. 256, No. 5519 (Aug. 21, 1975), pp. 608–10.

DERNBERGER, R. F., The transfer of technology to China, *Asia Quarterly,* Vol. 3 or 4 (1974), pp. 229–52.

DERNBERGER, R. F., *Economic Development and Modernization in Contemporary China. Statistical Appendix,* forthcoming in *Technology and Communist Culture,* Praeger, New York.

*Directory of Scientific Research Institutes in the People's Republic of China,* by S. SWANNACK-NUNN: Vol. 1, *Agriculture, Fisheries, Forestry.* Vol. 2, *Chemicals, Construction.* Vol. 3, Part I, *Electrical and Electronics, Energy, Light Industry, Machinery Including Metals and Mining, Transportation.* Vol. 3, Part II, *Light Industry, Machinery Including Metals and Mining, Transportation,* The National Council for US–China Trade, Washington, DC, 1977–8.

*Directory of selected scientific institutions in Mainland China,* prepared by Surveys & Research Corporation (R. J. Watkins, ed.), Hoover Institution Publications Series, 96, Hoover Institutions Press, Stanford, Calif., 1970, 469 pp.

DONNITHORNE, A., *China's Economic System,* Praeger, New York, 1967.

*Earthquake Research in China,* report by the American Seismological Delegation, Oct. 5 – Nov. 5, 1974, *EOS,* Vol. 56, No. 11 (1975), pp. 838–81.

ECKSTEIN, A., *China's Economic Revolution,* Cambridge University Press, New York, 1977.

ESPOSITO, B. J., The politics of medicine in the People's Republic of China, *Bulletin of the Atomic Scientists,* Dec. 1972, pp. 4–9.

ESPOSITO, B. J., Science in Mainland China, *Bulletin of the Atomic Scientists,* Jan. 1972, pp. 36–40.

FITZGERALD, A. and SLICHTER, C. P., (eds.), *Solid States Physics in China,* a trip report of the American Solid State Physics Delegation, CSCPRC Report Series 1, Washington, DC: National Academy of Sciences, 1976.

GANIERE, N., *The Process of Industrialization of China. Primary elements of an Analytical Bibliography,* Development Centre, OECD, Paris, 1974, 137 pp.

GOULD, S. H., (ed.), *Sciences in Communist China,* American Association for the Advancement of Science, Washington, 1961.

*Herbal Pharmacology in the People's Republic of China,* a trip report of the American Herbal Pharmacology Delegation, National Academy of Sciences, Washington, DC, 1975.

HEYMANN, Jr., H., *China's Approach to Technology Acquisition,* Part I, *The Aircraft Industry,* Rand Corporation, Santa Monica, Calif., 1975 (R-1573-ARPA), 75 pp.

HEYMANN, Jr., H., *China Approach to Technology Acquisition, Part III, Summary Observations,* A report prepared for Defense Advanced Research Projects Agency, Rand Corporation, Santa Monica, Calif., 1975 (R-1575-ARPA), 72 pp.

HEYMANN, Jr., H., *"Self-reliance" Revisited: China's Technology Dilemma,* 1974, Rand Corporation, 25 pp.

HORVATH, J., *Chinese Technology Transfer to the Third World – A Grants Economy Analysis,* Praeger, New York, 1976.

Insect control in the People's Republic of China (excerpted from "A note on the choice of technology in China"), *Journal of Development Studies,* Vol. IX, No. 1 (Oct. 1972), pp. 98–108.

ISHIKAWA, S., Chinese method for technological development – the case of agricultural machinery and implement industry, a paper presented at the OECD Symposium on Science, Technology, and Development in a Changing World, Paris, April 21–25, 1975, *The Developing Economies,* Vol. 13, No. 4 (Dec. 1975), pp. 430–57.

ISHIKAWA, S., Choice of techniques in mainland China, *The Developing Economies,* Sept.–Dec. 1972, pp. 24–56.

ISHIKAWA, S. and TAMURA, S. (eds.), *Chugoku ni okeru kagaku gijutsu hatten no ichi kenkyū (A Study on the Development of Chinese Science and Technology)*, Japan Economic Research Centre, Tokyo, 1975.

ISHIKAWA, S., Technological choice in China (excerpted from "A note on the choice of technology in China"), *Journal of Development Studies*, Vol. IX, No. 1 (Oct. 1972).

KAPP, K. W., *Environmental Policies and Development Planning in Contemporary China and other Essays*, Haag, 1974.

KEESING, D. B., Economic lessons from China, *Journal of Development Economics* Vol. 2 (1975), pp. 1–32.

KLOCHKO, M. A., *Soviet Scientist in Red China* (translated by Andrew MacAndrew), London, Hollis & Carter, 1964, 192 pp.

KOJIMA, R., *Chugoku no keizai to gjutsu (China's Science and Technology)*, Keizō shobō, Tokyo, 1975.

KUO, L. T. C., *Agriculture in the People's Republic of China. Structural Changes and Technical Transformation*, Praeger, New York, 1976, 288 pp.

LATEEF, A., *Economic Growth in China and India 1950–1980*, Vol. 1, *Report*, 96 pp.; Vol. 2, *Appendices*, 64 pp., London, 1976. EIU Special Report No. 30, The Economic Intelligence Unit Ltd.

LARDY, N. R., Economic planning and income distribution in China, *Current Science*, Vol. XIV, No. 11 (Nov. 1976), pp. 1–12.

LINDBECK, J. M. H., The organization and development of science, *The China Quarterly*, Apr.– June 1961, pp. 98–132.

LUBKIN, G., Physics in China, *Physics Today*, Dec. 1972, pp. 23–28.

McFARLANE, B., Mao's game plan for China's industrial development, *Innovation*, Aug. 23, 1971, 12 pp.

MACIOTI, M., Hands of the Chinese, *New Scientist and Science Journal*, Vol. 50 (June 10, 1971), pp. 636–9.

MACIOTI, M., Scientists go barefoot, *Successo*, Jan. 1971, pp. 115–18.

MARTIN, C. M., China: future of the university. *Bulletin of the Atomic Scientists*, Jan. 1971, pp. 11–15.

MARU, R., *Research and Development in India and China*, a comparative analysis of research statistics and research effort, Research Policy Programme, University of Lund and Centre for the Study of Developing Societies, Delhi, Lund, 1969, 71 pp. + app.

MOREHOUSE, W., Notes on Hua-tung Commune: the commune as a technological system, *The China Quarterly*, No. 67 (Sept. 1976), pp. 582–96.

NICKUM, J. E., *A Collective Approach to Water Resource Development: The Chinese Commune System, 1962–1972*, Center for Chinese Studies, University of California, Berkeley, Calif., 1974, 334 pp. + app.

NORMAN, M., China's approach to environmental conservation, *Environmental Affairs*, winter 1976.

NUNN, S. S., (see under *Directory*)

NUNN, S. S., Research institutes in the People's Republic of China, *US–China Business Review*, Vol. 3, No. 2, (Mar.–April 1976), pp. 39–50.

OECD COMMITTEE FOR SCIENTIFIC AND TECHNOLOGICAL POLICY, *Science and technology in the People's Republic of China*, Paris 1976 (SPT (77)1), OECD, 355 pp.

OLDHAM, C. H. G., Science and education, *Bulletin of the Atomic Scientists*, June 1966.

ORLEANS, L. A., China's science and technology: continuity and innovation, *People's Republic of China: An Economic Assessment*, US Congress, Washington, 1972, pp. 185–219.

ORLEANS, L. A., Communist China's education: policies, problems and prospects, *An Economic Profile of Mainland China*, Vol. II, US Congress, Washington, 1967, Vols. I & II, pp. 499–518.

ORLEANS, L. A. and SUTTMEIER, R., The Mao ethic and environmental quality, *Science*, Dec. 1970.

ORLEANS, L. A., Research and development in Communist China, *Science*, Vol. 157, (July 28, 1967), pp. 392–400.

ORLEANS, L. A., Research and development in communist China: Mood, management and measurement, *An Economic Profile in Mainland China*, US Congress, Joint Economic Committee, Washington DC, US GPO, 1967, pp. 549–78.

*Plant Studies in the People's Republic of China,* a trip report of the American Plant Studies Delegation, Washington, DC, National Academy of Sciences, 1975.

PRYBYLA, J. S., *The Political Economy of Communist China. International Textbook,* Scranton, Pa., 1970.

*Pure and Applied Mathematics in the People's Republic of China,* CSCPRC Report Series 3, National Academy of Sciences, Washington, DC.

RICHMAN, B. M., *Industrial Society in Communist China,* Random House, New York, 1969.

RIDLEY, C. P., *China's Scientific Policies. Implications for International Cooperation,* Washington, DC, 1976 (AEI-Hoover Policy Study 20, Hoover Institution Studies 50), American Enterprise Institute for Public Policy Research and Hoover Institution, Stanford, Calif., 92 pp.

RIFKIN, S. B., The Chinese model for science and technology: its relevance for other developing countries, *Development and Change,* Vol. 6, No. 1 (Jan. 1975), pp. 23–40.

ROBINSON, J., *Economic Management in China,* Modern China Series, No. 4, Anglo-Chinese Educational Institute, London, 1975.

ROYAL DANISH ACADEMY OF SCIENCES AND LETTERS, *Natural Sciences in China,* report from a visit to the PRC by a delegation from the RDASL, Copenhagen 1976, the Academy and the Danish Research Administration, 89 pp.

*Rural Small-scale Industry in the People's Republic of China,* by the American Rural Small-scale Industry Delegation, University of California Press, Berkeley, 1977.

SALAFF, S., *A Biography of Hua Lo-keng, in Science and Technology in East Asia* (Natan Sivin, ed.), Neale Watson Academic Publications, New York, 1977.

SCHURMANN, F., *Ideology and Organization in Communist China,* Berkeley, Los Angeles, London, 1970.

SATŌ, M., Technological development in China viewed through the electronics industry: an engineer's view, *The Developing Economies,* Tokyo (Sept. 1971), pp. 315–331.

SCHWARTZ, J., *Computing in China,* a trip report, June–July 1973, Courant Institute of Mathematical Sciences, New York University, New York, 1973 or 1974, 13 pp.

*Science and Technology in the People's Republic of China,* OECD, Paris, 1977.

SHAPLEY, D., Chinese science: what the China watchers watch, *Science,* Vol. 173, No. 3997, (Aug. 13, 1971), pp. 615–17.

SIGNER, E. and GALSTON, A. W., Education and science in China, *Science,* Vol. 175, (Jan., 1972), pp. 15–23.

SIGURDSON, J., *Naturvetenskap och teknik i Kina,* Stockholm, 1968 (IVA meddelande 154), Ingenjörsvetenskapsakademien, 82 pp.

SIGURDSON, J., Resources and Environment in China, *AMBIO* (Swedish Academy of Sciences, Stockholm), Vol. 4, No 3 (1975), pp. 112–19.

SIGURDSON, J., *Rural Industrialization in China.* Harvard University Press, Cambridge, 1977, 281 pp., illustr., bibliog., index.

SIGURDSON, J., *Small-scale Cement Plants – A Study in Economics,* Intermediate Technology Publications, London, 1977, 28 pp., illus., bibliog.

SIGURDSON, J., Technology and employment in China, *World Development,* Vol. 2, No 3 (Mar. 1974), pp. 75–85.

SIGURDSON, J., The suitability of technology in contemporary China, *Impact of Science on Society,* Vol. XXIII, No 4 (1973), pp. 341–51.

SMIL, V., *China's Energy,* Praeger, New York, 1977.

SMIL, V., Energy in China: achievements and prospects, *The China Quarterly,* Jan. 1976.

SMIL, V., Intermediate energy technology in China, *Bulletin of the Atomic Scientists,* Vol. 33, No 2 (Feb. 1977), pp. 25–31.

STAVIS, B., Agricultural Research and Extension: the Chinese Model. April 1977, Ithaca, NY, 1977, 112 pp. (mimeo. paper prepared for a conference, Chinese Rural Institutions: Lessons for Other Developing Countries, at IDS, University of Sussex, Brighton, England, Apr. 14–17, 1977), Cornell University.

STAVIS, B., *Making Green Revolution. The Politics of Agricultural Development in China,* Cornell, NY, 1974, 274 pp. (Rural Development Monograph No. 1), the Rural Development Committee, Cornell Univ. and China–Japan Program, Cornell Univ.

*Survey of PRC Literature on Science and Technology: A bibliography,* partly annotated by E. Baark, R. Jonsen, and D. Wagner, Research Policy Programme, Lund, Sweden, 1977, 46 pp.

SUTTMEIER, R. P., Recent developments in the politics of Chinese science, *Asian Survey,* Vol XVII, No. 4 (Apr. 1977), pp. 375–92.

SUTTMEIER, R. P., Record from the first decade – party views of science, *The China Quarterly*, No. 44 (Oct.–Dec. 1970), pp. 146–68.
SUTTMEIER, R. P., *Research and Revolution. Science Policy and Societal Change in China*, Lexington, Mass., 1974, Lexington Books, 188 pp.
SUTTMEIER, R. P., Science policy shifts: organization, change and China's development, *The China Quarterly*, No 62 (June 1975), pp. 207–41.
SWEDISH ACADEMY *Science, Technology in Communist China*, Report No. 154 by J. SIGURDSON, published by Swedish Academy of Engineering Sciences, Stockholm, Naturvetenskap och teknik i Kina (Natural Science and Technology in China), Swedish, 1968, pp. 1–82, JPRS/L 2972 (Feb. 28, 1969).

TIEN, H. Ti, China's institute of biophysics and other scientific institutions, *Eastern Horizon*, Vol. 13, No 5 (1974), pp. 39–55.
TSU, R., High technology in China, *Scientific American*, Vol. 227, No. 6 (Dec. 1972), pp. 13–17.

UCHIDA, G., Technology in China, *Scientific American*, Vol. 215, No. 5 (1966), pp. 37–45.
US CONGRESS, *China: A Reassessment of the Economy*, Washington DC, 1975 (a compendium of papers submitted to the Joint Economic Committee, Congress of the US, 94th Congress, 1st Session), US GPO, 737 pp.
US CONGRESS, *People's Republic of China: An Economic Assessment*, Joint Economic Committee. Washington DC, 1972 (May 18, 1972), US GPO, 382 pp.
US CONGRESS, Chinese Economy Post-Mao, A compendium of papers submitted to the Joint Economic Committee, Congress of United States, Volume 1. Policy and Performance, Washington, D.C. 1978, 880 pp.
US HOUSE OF REPRESENTATIVES, Annotated bibliography on science and technology in China. Science and technology in the People's Republic of China.

WANG, K. P., *Mineral Resources and Basic Industries in the People's Republic of China*, Westview Press, Boulder, Colorado, 1977.
WANG, K. P., *The People's Republic of China. A New Industrial Power with a Strong Mineral Base*, US Bureau of Mines, Washington, 1975, US GPO, 96 pp. + map.
WHEELWRIGHT, E. L. and MCFARLANE, B., *The Chinese Road to Socialism; Economics of the Cultural Revolution*, New York and London, 1970.
WHITNEY, J. B. R., Ecology and environmental control in China, in *China's Developmental Experience* (M. Oksenberg, ed.), Praeger, New York, 1974.
WORTMAN, S., Agriculture in China, *Scientific American*, Vol. 232, No. 6 (June 1975), pp. 13–21.
WU, YUAN-LI and SHEEKS, R. B., *The Organization and Support of Scientific Research and Development in Mainland China*, New York, 1970, Praeger Special Studies in International Economics and Development, published for the National Science Foundation, Praeger, New York, 592 pp.

VYLDER, S. de, *Foreign Trade and Self-reliance in China. An introduction*, Stockholm, 1974 (China's Developmental Strategy, No. 11), EFI, the Economic Research Institute at the Stockholm School of Economics, 81 pp.

YAMADA, K., The Development of Science and Technology in China: 1949–65, *The Developing Economics*, No. 4 (1970), pp. 503–37.
YANG, C. N., C. N. Yang Discusses Physics in People's Republic of China, *Physics Today*, Nov. 1971, pp. 61–63.

# Index

Academy of Agriculture and Forestry Sciences 57

Academy of Sciences (Academia Sinica) 5n, 12, 14, 15, 42, 46, 47, 49, 58, 61–6, 68, 69, 72, 73, 82, 89, 91, 93, 100, 113, 116

Academy of Social Sciences 14, 54, 93, 113

Aeronautics 38

Agricultural science 102, 104, 105–6

Amateur scientists *see* Mass science

Aquaculture 116

Atmospheric Physics, Institute of 113

Atomic Energy Institute 68, 68n, 69

Autarky *see* Self-sufficiency

Awards (for inventors) 13

Baark, Erik 55

Barefoot doctors 108

Basic research 95–101

Beijing Aeronautical Engineering Institute 87

Beijing University 50, 83, 87

Billgren, Boel 68n

Biogas 116

Biological pest control 116–17 ·

Biosciences 62, 63

Brown, Shannon 42

Cancer 108–9
  Control Office, National 108

Cardettini, Onelia 35, 37, 39

Catching up *see* Technology gap

Chaoyang Agricultural College 101

Chemistry, Institute of 113

Class struggle 99

Coal 26, 38, 39, 42, 116

Collaboration in science and technology 15–16; *see also* Joint ventures

Collaboration with USSR 4, 31

Colleges *see* Universities

College Enrolment, National Conference on 89; *see also* Universities

Commissions *see under* State

Commune- and brigade-run enterprises *see* Small-scale industry

Computers 38, 130, 134, 137; *see also* Electronics

Conference, National 13–14

Confucius 98

Cooper, Julian 137, 139

Cultural Revolution 1, 5, 16, 19, 22, 42, 48, 50, 51, 52, 58, 62, 81, 82, 86, 90, 92, 99

Daqing 109

Daqing Petroleum Institute 87

Dazhai Agricultural College 87

Decentralization of research 100–1; *see also* Regional science

Defence technology *see* Military technology

Deng Xiaoping 5, 15, 40, 51

Dernberger, Robert 139

Dongji University 87

Dual economy 23; *see also* Two legs and Small-scale industry

Earthquake production *see* Seismology

Ecological system 112; *see also* Environmental production

Education 13
  Conference, National 74, 77
  programme, national 74
  Ministry of 13, 58, 62, 81, 86, 89
  Ministry of Higher 58
  educated young people (resettlement of) 80, 103

Electronics 41, 42, 124–36
  components 128
  Industry Conference, National 124, 129–31
  integrated circuits 38, 128, 131

Elzinga, Aant 54

Energy 116, 119–22
  consumption 119
  electricity generation 39
  hydropower 119

Engineering manpower *see* Manpower

Engineers *see* Manpower

Environmental Protection 112–19
  environment, Stockholm Conference on the 113
  Environmental Protection Office 113, 118
  night soil 113, 115

European Economic Community, trade agreements with 43

Fang Yi 12, 14, 27n, 87

Finance and Trade, National Conference on 4

Five-Year Plan
  Third 2
  Fifth 88
  Sixth 2, 88, 131

Foreign technology 1, 4–5, 29, 35–6, 43, 44
  import of complete plants 29; *see also* Technology Transfer

167